Concepts and Principles of Geochemistry

Concepts and Principles of Geochemistry

Edited by **Karolina Jensen**

New York

Published by Callisto Reference,
106 Park Avenue, Suite 200,
New York, NY 10016, USA
www.callistoreference.com

Concepts and Principles of Geochemistry
Edited by Karolina Jensen

International Standard Book Number: 978-1-63239-126-1 (Hardback)

Printed in the United States of America.

Contents

Preface

This book was inspired by the evolution of our times; to answer the curiosity of inquisitive minds. Many developments have occurred across the globe in the recent past which has transformed the progress in the field.

The book consists of information regarding several issues within the area of geochemistry. The audience for this book includes a large number of scientists like physicists, geologists, geochemists, technologists and government agencies. The issues introduced establish a starting point for new ideas and further additions. An effective and efficient management of geological and environmental topics necessitates the understanding of current research in the field of soil, ores, rocks, water, and sediments. The research that is showcased in the book was accomplished by experts and is thus suggested to scientists, under and postgraduate students who want to acquire knowledge and insight into the latest advancements in geochemistry and also benefit from an improved understanding of the earth's system processes.

This book was developed from a mere concept to drafts to chapters and finally compiled together as a complete text to benefit the readers across all nations. To ensure the quality of the content we instilled two significant steps in our procedure. The first was to appoint an editorial team that would verify the data and statistics provided in the book and also select the most appropriate and valuable contributions from the plentiful contributions we received from authors worldwide. The next step was to appoint an expert of the topic as the Editor-in-Chief, who would head the project and finally make the necessary amendments and modifications to make the text reader-friendly. I was then commissioned to examine all the material to present the topics in the most comprehensible and productive format.

I would like to take this opportunity to thank all the contributing authors who were supportive enough to contribute their time and knowledge to this project. I also wish to convey my regards to my family who have been extremely supportive during the entire project.

Editor

Trace Metals in Shallow Marine Sediments from the Ría de Vigo: Sources, Pollution, Speciation and Early Diagenesis

Paula Álvarez-Iglesias and Belén Rubio
Universidad de Vigo, Vigo (Pontevedra)
Spain

1. Introduction

The maintenance of the environmental quality of coastal environments requires a good knowledge on them, identifying their potential problems and considering the different anthropogenic activities that take place in these settings. A great proportion of the suspended material that arrives to these environments is incorporated to the bottom sediments. Sedimentation is favoured due to the abrupt changes that are produced in the physico-chemical parameters (pH, Eh, salinity, etc.) by the confluence of continental and marine waters. Their sediments constitute environmental archives, recording trace element inputs to the marine environment. Metal concentrations in these sediments usually surpass in several orders of magnitude those existing in the adjacent water column and in the interstitial waters (Tessier & Campbell, 1988). Their analysis allows covering different objectives, such as studying the spatial and temporal history of pollution of a particular place (Zwolsman et al., 1993), detecting pollutant sources (Dassenakis et al., 1996) and evaluating their potential effects to the organisms (Fichet et al., 1998). Sediments can act as a secondary source of pollution by resuspension or dissolution (early diagenesis) processes. Their study needs a multidisciplinary approach, taking into account the different physico-chemical processes operating in the study area and considering the relationships between metal concentration, mineralogy, grain-size and metal sources. All these variables are reviewed in this chapter, showing examples of sediments from a shallow transitional environment: the Ría de Vigo (NW Spain).

The Ría de Vigo is the sourthernmost ría of the Galician Rías Baixas (Fig. 1) with approximately 30 km in length. At its mouth the Cíes Islands are located, which act as a barrier against storms. It can be differentiated the outer and middle ría sectors from the inner sector of the ría (the Bay of San Simón). The central axis of the former runs NE-SW, while the axis of the inner sector is NNE-SSW oriented. This last sector, separated from the rest of the Ría by the Rande Strait (depth > 30 m), is very shallow (average depth of 7 m, in front of ~20 m in the remaining ría). It is a low energy area, where hydrodynamic conditions are ruled by tides, whereas water circulation in the middle and outer ría sectors is conditioned by waves and littoral drift and exhibits a two-layered positive residual circulation pattern (Álvarez-Salgado et al., 2000). Rivers are relatively small and mostly discharge in the sourthern margin and in the inner sector (Pérez-Arlucea et al., 2005). They run on a watershed mainly composed by granites, schist and gneiss (IGME, 1981). Surface sediment distribution is conditioned by hydrodynamic conditions: coarse-grained particles of fluvial origin concentrate close to the river inputs, whilst fine-grained particles are located in the central axis and in the inner sector of the ría, and biogenic sands are located at the northern margin of the outer ría (Vilas et al., 2005). The area is subjected to seasonal processes of upwelling/downwelling (Álvarez-Salgado et al., 2000). Upwelling events contribute to the ecological richness of the ría, where marine organisms (fish, shellfish) are exploited, some of them in rafts (floating platforms) (Freire & García-Allut, 2000). This high productivity explains the high organic matter contents observed in the ría sediments (Rubio et al., 2000a; Vilas et al., 2005), which are maxima toward the inner ría (up to 15%; Álvarez-Iglesias et al., 2003, 2006, 2007) and are positively correlated with the fine-grained particles content. Carbonate content is inversely correlated with these variables. Its percentage is negligible in inner ría sediments (Álvarez-Iglesias et al., 2003) and maximum towards the ría mouth (Rubio et al., 2000a). The main industrial activities in the ría are shipbuilding, automobile manufacture, ceramic manufacture and canned food (González-Pérez & Pérez-González, 2003). The Vigo harbour, which concentrates many of the industrial activities, is located in the middle ría sector, and the wastewater plant of the city of Vigo (the biggest settlement in the ría), is located in the middle-outer ría sector, close to the mouth of the Lagares River (Fig. 1).

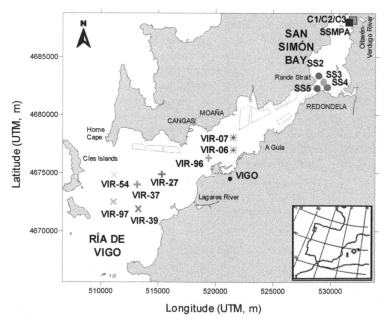

Fig. 1. Study area. Location of sampling points considered in the present study: inner ría (squares and circles), middle ría (doble crosses), middle-outer ría (crosses) and outer ría (vanes) cores. Note that location for intertidal cores C1, C2 and C3 is showed by a single symbol due to the map scale, although they have been recovered at different points. Polygonal areas represent current mussel raft areas.

2. Sources of metals to coastal sedimentary environments

Metal content in marine sediments reflects both natural and anthropogenic components. Inputs from weathering of soils, rocks and ores from the watershed, constitutes the lithogenic natural component (background level, see subsection 5.1 for details). The natural component has also inputs from biogenic production, that are going to dilute both lithogenic and anthropogenic inputs. The anthropogenic component is constituted by those metals released to the environment as a consequence of different human activities.

Metal inputs to the marine environment can be direct or indirect. Direct inputs come from urban sewage, industrial wastes (metallic, chemical, building and shipping industries) and metal production and recycling. Indirect inputs come from atmosphere and from rivers as dissolved or suspended load (Salomons & Förstner, 1984). Most of the inputs consist of a group of trace elements (As, Cd, Cu, Co, Fe, Mn, Pb and Zn, among others) usually known as heavy metals. This term, according to Moore (1991) refers to a group of persistent metals and semimetals with relatively high density and that are usually toxic or poisoinous at low concentrations. Nevertheless, according to Duffus (2002), the term heavy metal has not a precise deffinition, it is outdated and it should be substituted for another more adequate such as trace element or trace metal depending on the context.

Once trace elements arrive to the marine environment they are going to be distributed between the different compartments by association with dissolved organic and inorganic ligands or with particulate matter, and by ingestion by organisms.

In the particular case of the Ría de Vigo some metal enrichments have been detected for particular areas in several works on surficial sediments: around the harbour area (Cu, Pb, Zn, Fe), the outflow of the Lagares River (Pb, Zn) and the inner ría sector (Pb) (for details see the review by Prego & Cobelo, 2003). Metal ranges of concentrations and metal speciation will be considered in the following sections.

3. Metal concentration *vs* grain size

Trace metals in the marine environment are usually associated to organic-rich fine-grained sediments (Förstner, 1989) related to the good complexation and peptidizing properties of organic matter (Wangersky, 1986) and the high specific surface and surficial charge of clay minerals (Horowitz & Elrick, 1987). The presence of coarse-grained particles in the sediments, that are mostly composed by inert substances (silicates, carbonates, feldspars), usually cause trace metal dilution (Förstner, 1989).

The grain-size effect can be observed in Ría de Vigo sediments, where elements associated with fine-grained fractions, such as Al, V or Ni, are correlated with the mud fraction of sediments (Fig. 2). In opposition, elements such as Si are more abundant in coarse-grained sediments (Fig. 2). Elements such as Ca and Sr, with a biogenic origin are mainly associated with the coarse-grained fractions, and cause a dilution of another elements contents (Fig. 2).

3.1 Normalization procedures

Several normalization methods are generally used in order to correct for grain-size effects (Fig. 3). They are based in different approaches: 1) Indirect approximations (Summers et al., 1996); 2) The physical separation of a specific grain-size fraction –usually that <63 μm

(Förstner, 1989); and/or 3) The normalization of the metal concentrations by the content of a conservative element associated to the fine-grained fraction. In the last case Al, Fe, Li and Rb are usually selected (Ackerman, 1980; Álvarez-Iglesias et al., 2003, 2006; Loring, 1990). Lithium is preferred in front of Al when analyzing sediments from high latitudes (Loring, 1990). Some workers (Prego et al., 2006, among others) have chosen Fe for normalizing because its high abundance in the Earth's crust and its similar behavior and geochemistry to those of many trace elements but this metal shows a high mobility by postdepositional diagenetical processes (Álvarez-Iglesias et al., 2003; Rey et al., 2005) and, furthermore, different coastal areas can show Fe pollution (Rubio et al., 2000a).

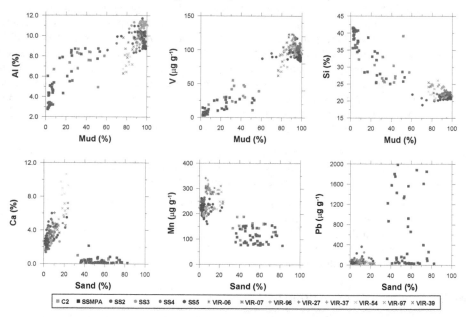

Fig. 2. Scatterplots of element concentrations vs mud (<63 μm) or sand contents in core samples (N = 210) from different sectors of the Ría de Vigo (core depths between 0.24 and 3.10 m): inner ría both intertidal (squares) and subtidal (circles) sediments, middle ría (doble crosses), middle-outer ría (crosses) and outer ría (vanes). Metal concentrations obtained from X-ray fluorescence (XRF) analyses. Note the positive correlation between Al or V vs mud and Ca vs sand, and the negative correlation between Si vs mud and Mn vs sand contents. Note also the extremely high values of Pb in the intertidal sediments. Sample location in Fig. 1.

The approaches 2 and 3 are generally combined (Álvarez-Iglesias et al., 2003) because, even in the grain-size fraction <63 μm, metal contents are not homogeneously distributed (Salomons et al., 1985). Nevertheless, in studies where fine-grained content is very high, normalization is usually accomplished by the use of a normalizer element on bulk sample results. These grain-size effects on metal concentrations can be clearly observed in the scatterplots of Fig. 3, where concentrations of Cr, Cu and Zn in Ría de Vigo sediments increased with the Al content. Note also in Fig. 3 that some samples show metal enrichments lying out of the general trend. This will be discussed in the following sections.

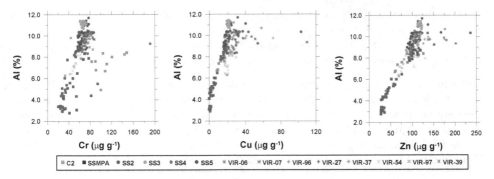

Fig. 3. Scatterplots of Al vs Cr, Al vs Cu and Al vs Zn in core samples (N=210) from different sectors of the Ría de Vigo (sample location in Fig. 1). Metal contents obtained from XRF analyses.

4. Determination of total metal concentration and speciation

The different anthropogenic inputs of metals to the marine environment will cause changes in sediment composition. A detailed study on the sedimentary record will allow evaluating environmental changes both spatially and temporally for a particular coastal area. It is important to consider also metal speciation, to evaluate potential mobility and bioavailability of the different trace metals (Bryan & Langston, 1992).

4.1 Total metal concentration

Total metal concentrations can be determined by XRF or by measuring metal contents, after acid digestion of the sediments, by different techniques such as Atomic Absorption Spectrometry (AAS) or Inductively Coupled Plasma-Mass Spectrometry (ICP-MS), among others.

Advantages of XRF are: 1) easy sample preparation, 2) quickness, 3) relatively low cost of the measurements, and 4) high analytical resolution (Boyle, 2000). Continuous high-resolution XRF core scanner devices allow making a qualitative exploratory analysis very quickly (1.0 m of a core in a few hours) analyzing directly fresh and undisturbed sediments (Rubio et al., 2011). This analysis can be refined by analyzing discrete samples by tradicional XRF techniques (obtaining quatitative data).

Total metal concentrations can be also determined by digestion of the samples in open or closed systems with heating. The different workers usually utilize the following reactants, alone or in combination: HCl, HNO$_3$, H$_2$O$_2$, H$_2$SO$_4$, HClO$_4$ and HF (Álvarez-Iglesias et al., 2003; Belzunce-Segarra et al., 1997; Varekamp, 1991). Traditionally samples were acid digested applying dry (oven, hot plate) or humid (water bath) heat, but in the last decades the use of microwave ovens has been extended (Álvarez-Iglesias et al., 2003; Izquierdo et al., 1997; Rubio et al., 2001). The use of a closed system has several advantages: 1) the attack is effective, 2) the analytical results are similar to those traditional (Mahan et al., 1987), 3) the human errors are minimized, 4) the analytical contamination problems are reduced, 5) the losses by volatilization are minimal, 6) the efficiency of destruction under pressure is high (Uhrberg, 1982), and 7) the time for analysis is reduced.

If acids such as HCl, HNO_3 or $HClO_4$ are used alone or in combination but not HF, the obtained metal contents do not represent total metal but pseudo-total metal concentrations (Barreiro Lozano, 1991). Then, the selection of one or another analytical technique for determining metal concentration will depend on the objective of the study.

In order to check for accuracy and precission of the results of metal concentration it must be analyzed reference materials by the same techniques that those applied in the sediments under study (Table 1). Furthermore, the reference materials should have similar characteristics than those of the target sediments to avoid the matrix effect (Boyle, 2000). In the case of the Ría de Vigo the reference materials usually analyzed have been MESS-1, MESS-3 and PACS-1 (Álvarez-Iglesias et al., 2003, 2006; Prego et al., 2006; Vega et al., 2008).

	MESS-1			PACS-1	PACS-1	MESS-3	MESS-3
	CV	Measured[1a]	Measured[2b]	CV	Measured[3]	CV	Measured[4d]
Al (%)	5.84±0.20	3.71±2.14	6.09	-	-	-	-
Fe (%)	3.05±0.17	2.47±0.12	2.99	4.87±0.08	5.11±0.09[a]	-	-
Co ($\mu g\ g^{-1}$)	10.8±1.9	-	9.4	-	-	-	-
Cr ($\mu g\ g^{-1}$)	71±11	-	74.7	113±8	99±10[c]	-	-
Cu ($\mu g\ g^{-1}$)	25.1±3.8	23.84±2.17	24.9	452±16	424±37[c]	33.9±1.6	31±5
Mn ($\mu g\ g^{-1}$)	513±25	403.5±35.2	451	-	-	-	-
Ni ($\mu g\ g^{-1}$)	29.5±2.7	-	29.2	-	-	-	-
Pb ($\mu g\ g^{-1}$)	34.0±6.1	36.50±7.24	38	404±20	409±24[c]	21.1±0.7	23.6±2.5
V ($\mu g\ g^{-1}$)	72.4±17	-	101.3	-	-	-	-
Zn ($\mu g\ g^{-1}$)	191.0±17.0	161.5±15.4	190.9	-	-	159±8	161±5

[1]Álvarez-Iglesias et al. (2003); [2]Álvarez-Iglesias et al. (2006); [3] Prego et al., 2006; [4]Vega et al. (2008)
[a]AAS; [b]XRF; [c]ET-AAS; [d]ICP-OES.

Table 1. Comparison of certified values and measured values for several reference materials obtained in different studies. CV: Certified value. Measured values were obtained by different techniques: analysis of sediment extracts from total acid digestion in microwave oven by AAS, analysis of bulk sediments by XRF, Electrothermal atomic absorption spectrometry (ET-AAS) and Inductively coupled plasma-optical emission (ICP-OES). Note the differences in metal concentrations obtained by different techniques for the reference material MESS-1.

4.2 Speciation

The bioavailability, toxicity and mobility of a particular trace element depends on the sedimentary phase to which is bound to, that is, on metal speciation. It is necessary to distinguish and quantify the different metal species to study current and potential impacts of polluted sediments, to identify metal sources, and to understand the geochemical processes that take place in a particular environment. Metal association to the sediments can be diverse: adsorbed onto surfaces of clay-particles, Fe-Mn oxyhydroxides and organic matter; occluded in amorphous materials, such as Fe-Mn oxyhydroxides, Fe sulphides or remanents of biological organisms; or being part of the mineral structure. Direct determination of each single chemical form of a metal is not practical, and indirect methods are preferred such as: statistical correlations, thermodynamical calculations based on models of chemical equilibrium, computer simulations, and more commonly chemical extraction methods (Luoma & Bryan, 1981; Luoma, 1986; Tessier et al., 1979; Tessier & Campbell, 1988).

Chemical methods can be selective or non-selective. Non-selective methods extract operationally defined metal fractions that are not directly identifiable with an specific phase in the sediment (Loring, 1981). Although they do not correspond to one isolated chemical form, they suppose a better estimation of reactivity instead of total metal content. These methods can be one-single step or sequentials. The first ones can be grouped in three categories depending on the selected reactant: 1) diluted solutions from strong acids; 2) weak acids; and 3) solutions from strong reducing or complexing agents (Bryan & Langston, 1992; Imperato et al., 2003). Advantages of these methods are quickness, simplicity, cost-efficiency relationship, allowance of a better contrast between background values and anomalous values, and the non-utilization of dangerous acids such as $HClO_4$ or HF. The main inconvenient is the difficulty of finding a single reactant with the capacity of extracting the labile metal fraction (or any other target fraction) without attacking the residual fraction.

Operationally defined fractions						Reference
A	B	C		D	E	Tessier et al., 1979
A+B		C+D			E*	Salomons & Föstner, 1984
A	B	C1	C2	D	E*	Salomons & Föstner, 1984
A	B	C1	C2	D	-	Kersten & Föstner, 1987
A+B+C			D1	D2	E*	Huerta-Díaz & Morse, 1990
A+B		C		D	-	Ure et al., 1993
A	B	C1	C2	D	E*	Borovec, 1996
A*	B+C*			D*	E	Dassenakis et al., 1996
A	B	C1	C2	D	-	Howard & Shu, 1996
A	B	C1	C2	D	E*	Izquierdo et al., 1997
A*	B'	C1'	C2'	D	-	Flyhammamar, 1998
A+B		C		D	E	Gómez-Ariza et al., 2000
A+B		C		D	E*	Stephens et al., 2001

A: exchangeable
B: bound to carbonates
C: bound to Fe-Mn oxyhydroxides (reducible)
D: bound to organic matter-sulphides (oxidizable)
E: residual
oxyhydroxides
A*: easily interchangeable
B + C*: non-exchangeable inorganic fraction
D*: organic fraction

C1: Mn oxides(easily reducible)
C2: Fe oxides (moderatelly reducible)
B': metals non-extracted in A* and bound to carbonates
C1': Mn oxides and amorphous Fe oxides
C2': amorphous and poor-cristalyzed Fe
D1: bound to organic matter
D2: bound to sulphides
E*: bound to silicates (but without using HF)

Table 2. Comparison between the different metal fractions obtained with different sequential extraction methods.

Sequential methods consist of extracting the same sediment sample with a sequence of reactants, usually among 3 and 8, with increasing reactivity in the disolution process or with different nature regarding the previous one. The application of these methods results in a series of operationally-defined phases. Reactants are diverse and can be grouped in: concentrated inert electrolites, weak acids, reducing agents, complexing agents, oxidizing agents and strong mineral acids. There are different extraction techniques (Table 2). The lack of uniformity between the different sequential extraction methods prevents comparison of the results among different techniques. The most applied sequential procedure is that of Tessier et al. (1979), directly applied, or with small modifications (Álvarez-Iglesias et al., 2003; Mahan et al., 1987).

The main inconvenients of sequential extraction methods are the non-specificity of the reactants, the redistribution of trace elements between phases and the overload of the chemical system if the metal content is too high. Furthermore, it has to be beard in mind that some speciation changes could occur because of sample preparation, and also, the lack of reference materials suitable for most of the utilized methods (Quevauviller et al., 1997).

Several complementary analysis can be performed to check for efficiency and selectivity of the extraction procedure: to analyze total organic and inorganic carbon and sulphide contents in the extracts or in the sediments that remains after the different extractions; to study the solids that remain after the different steps by XRF or Scanning Electron Microscopy (SEM); to extract pure samples of geochemically known phases; or to statistically analyze the metals extracted from the sediments and the main sedimentary components that can bind them (Luoma & Bryan, 1981; Rapin & Förstner, 1983).

Metal redistribution or readsorption between the remaining sedimentary phases during the sequential procedure is an inconvenient very difficult to avoid, this is due to the partial selectivity of the extractants (Tessier & Campbell, 1987). One example could be the existence of free sulphides that would cause precipitation of metals previously released from the sediments; another one would be sulphides oxidation (Ngiam & Lim, 2001). Several workers have shown that readsorption happens after the extraction (Guy et al., 1978; Rendell et al., 1980) but they have not agreed if this problem is significant or not.

The problems related to the overload of the chemical system can be solved repeating the same step several times. When metal content is high the quantity of reactant used can be insufficient to completely extract all the metal bound to a certain specific fraction, then, some of the metal would be still retained and would be extracted in the following step.

Regarding to the problems related to the possible changes in especiation because of sample preparation, several workers pointed out that this is more important in anoxic samples than in oxic samples, because most of the extraction procedures have been designed for oxic sediments (Tessier & Campbell, 1988). Techniques of sample preservation before analysis, frozen or keeping at low temperature are better preferred than drying.

In spite of the indicated inconvenients of sequential extraction procedures, these methods are useful to evaluate bioavailabity and potential mobility, and to study early diagenetical processes. Among them, the usually applied methods are those proposed by Tessier et al. (1979), Ure et al. (1993) and Huerta-Díaz & Morse (1990). In the Tessier et al. (1979) method (called in the following Tessier's method) the parameters involved in the solubilization of the extracted metals are carefully controled, and five operationally-defined fractions are obtained (Table 2; Fig. 4). The Ure et al. (1993) method, usually known as the BCR method separates three fractions (Table 2), whereas the Huerta-Díaz & Morse (1990) method (called in the following H&M's method) separates four fractions (Table 2, Fig. 4). One important difference is that with the last method metals bound to organic matter are differentiated from those metals bound to sulphides. Nevertheless, the Tessier's method allows differentiating between interchangeable, bound to carbonates and bound to Fe-Mn oxyhydroxides metal fractions, whereas all together represents one single fraction together with acid-volatile sulphides in the H&M's method and they are considered as two fractions in the BCR procedure (Table 2).

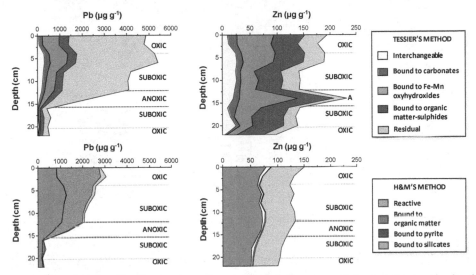

Fig. 4. Metal speciation distribution (μg g^{-1}) according to the sequential extraction methods of Tessier et al. (1979) and Huerta-Díaz & Morse (1990) for Pb and Zn in the fine-grained fraction of muddy sandy sediments (core C2) from inner Ría de Vigo. Note the different fraction recovery obtained for the same metal by the two procedures. Differences in the metal content in the silicate fraction or residual are due to the non-total leaching of silicates with the Huerta-Díaz & Morse method. Redox conditions (oxic, suboxic, anoxic-non sulphidic) in these sediments have been inferred according to DOP values (Álvarez-Iglesias et al., 2003).

The interchangeable fraction corresponds to those metals bound to low energy adsorption places and it is the most active fraction from a biological point of view (Campbell et al., 1988). In the fine-grained fraction of inner Ría de Vigo sediments it represents a relatively low content (average ~50 μg g^{-1} for Pb and ~10 μg g^{-1} for Zn, N–34), being more or less abundant depending on the grain-size distribution, the muddier the sediments, the higher the metal contents. The fraction bound to carbonates represents those metals bound or adsorbed into carbonates. In the fine-grained fraction of inner ría sediments it represents a relatively low content (average ~77 μg g^{-1} for Pb and ~12 μg g^{-1} for Zn, N=34). These two fractions together are more or less equivalent to the labile fraction of the BCR method. The labile fraction also represents a very low contribution in middle and outer ría sediments (<1.1 μg g^{-1} for Cu, non-detected for Pb and ~4.1 μg g^{-1} for Zn, N=47; Fig. 5).

The fraction bound to Fe-Mn oxyhydroxides corresponds to that metallic fraction bound to reducible Al, Fe and Mn oxides/hydroxides. They are unstable compounds under reducing conditions, which causes the release of associated metals. In the sediments of inner Ría de Vigo this fraction represents a significant percentage of the total concentration: around 20% of total Pb and 36% of total Zn (Fig. 4) decreasing to ~6.9% of total Pb and ~14% of total Zn in middle and outer ría sediments (Fig. 5).

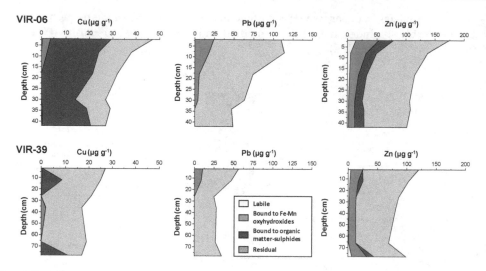

Fig. 5. Depthwise profiles of BCR fractions of Cu, Pb and Zn in cores from middle and outer ría (VIR-06 and VIR-39).

The metal fraction bound to carbonates, to amorphous and cristaline Fe-Mn oxyhydroxides (except goethite) and acid volatile sulphides (AVS) is extracted with the first step of the H&M's method, and then embracing the three fractions formerly discussed. It is called the reactive fraction. This fraction represents 43-75% of total Cu, 27-84% of total Pb and 33-57% of total Zn, in the fine-grained fraction of intertidal inner ría sediments (Fig. 4) and 0.1-51% of total Cu, 54-94% of total Pb and 11-53% of total Zn in subtidal inner ría sediments (Fig. 5).

The metal fraction bound to organic matter can be a significant sink for several metals depending on their relative abundance related to any other sedimentary phases. For example, this fraction represents 1.2-15% of total extracted Cu, 0.2-2.0% of total extracted Pb and 1.7-6.0% of total extracted Zn in intertidal and subtidal oxic sediments (N=20) of inner Ría de Vigo (Fig. 6), being their contribution lower in suboxic and anoxic sediments of the ría. Degradation of organic matter under oxidizing conditions will cause the release of metals bound to this phase.

The metal fraction bound to sulphides is significant in anoxic sediments, where metal sulphides, mainly pyrite, constitute a significant sink for several trace elements, as observed in the Ría de Vigo (Figs. 4, 6). This fraction can represent a high percentage of the total concentration of certain trace elements, such as Cu. For example, this fraction represents 26-94% of total extracted Cu, 5.0-47% of total extracted Pb and 1.2-9.4% of total extracted Zn in anoxic sediments (N=47) of inner Ría de Vigo, with a higher contribution for Cu and Zn in subtidal (Fig. 6) than in intertidal sediments.

These two fractions (bound to organic matter and to sulphides) are obtained as one single fraction, called oxidizable fraction, when applying the Tessier's and the BCR methods. The oxidizable fraction in the fine-grained fraction of intertidal sediments of inner Ría de Vigo according to the Tessier's method represents 7.4-31% of total extracted Pb and 6.1-51% of total extracted Zn (N=34, see examples for particular cores in Fig. 4), whereas this fraction in middle and outer ría sediments represents, on average, 46% of total extracted Cu, ~1.0% of total extracted Pb and ~11% of total extracted Zn (N=47) in middle and outer ría sediments according to the BCR procedure (see examples for particular cores in Fig. 5).

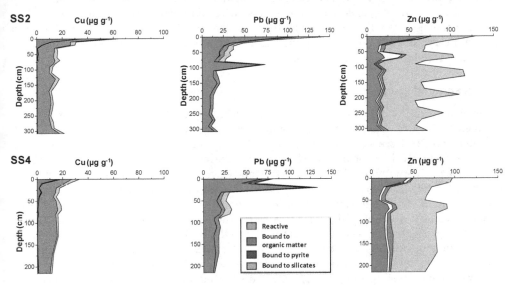

Fig. 6. Metal speciation distribution (µg g⁻¹) for Cu, Pb and Zn according to the Huerta-Díaz & Morse (1990) method for subtidal sediments of inner Ría de Vigo (cores SS2 and SS4).

The residual fraction is composed of primary and secondary detrital silicates that occur naturally and can hold trace elements into their crystal structure, resistant sulphides, and refractory organic matter (Tessier et al., 1979). Under natural conditions trace elements bound to this fraction are not expected to be released to the environment. Metal content in this fraction depends on mineral composition of the watershed, on physical properties that rules transport and deposit (grain-size, density, etc.) and on mineral stability and weathering intensity. This fraction is not obtained by some sequential procedures (Table 2) and then it can be obtained by subtraction. In other sequential methods, a silicate bound fraction is obtained but it does not represent the whole metal bound to this fraction. For example, in Fig. 4 the comparison of total concentrations obtained by the Tessier's and H&M's procedures showed clearly that the recovery for residual Pb was not complete with the second method. The comparison of total metal concentrations with the total recovery after applying these sequential extraction procedures showed differences depending on grain-size and mineralogy (Álvarez-Iglesias & Rubio, 2008, 2009): metal recovery was around 60-100% for Fe, 70-75% for Cu, 46-55% for Pb and 70-100% for Zn in inner ría sediments both subtidal and intertidal (fine-grained fraction). The residual fraction in intertidal sediments of inner Ría de Vigo according to the Tessier's method represents ~56% of total extracted Pb and ~18% of total extracted Zn (N=34, see examples in Fig. 4), whereas this fraction in middle and outer ría sediments represents on average ~52% of total extracted Cu, ~92% of total extracted Pb and ~71% of total extracted Zn (N=47) in middle and outer ría sediments according to the BCR procedure (see examples in Fig. 5).

5. Assessment of pollution in marine sediments

Establishing the degree of pollution by trace elements in marine sediments allows studying pollutants distribution for a particular area spatially and temporally and identifying pollutant sources (Dassenakis et al., 1996; Zwolsman et al., 1993). This is usually approached by determining several indexes and studying the relationships between metals and sediment characteristics. First, background levels (BLs) have to be considered, and second, the degree of pollution can be evaluated.

5.1 Background levels

Metal BLs correspond to the natural metal concentrations that exist or would exist in sediments without human influence. Several authors have proposed different BLs to global, regional and local scales, summarized in Tables 3 and 4, for major, minor and trace elements, respectively. Global or world-wide BLs correspond to average values calculated for the Earth crust (Taylor, 1964), granites, deep-sea clay sediments or lutites (Turekian & Wedepohl, 1961). Regional and local BLs take into account local geochemical variability (Rubio et al., 2000a). They can be determined in sediments from the target area or from adjacent areas by different methods such as 1) determination of pre-industrial metallic levels in dated cores; 2) application of regresion techniques between a geochemically-stable element without human-influence and the rest of metals; 3) selection of pristine areas; 4) selection of the first-percentile in accumulated distributions of metal concentration; 5) determination of homogenous populations into the target data; 6) a combination of the previous procedure with modal analysis; 7) a direct comparison of polluted and non-polluted sediments from the same study area or an adjacent area; or 8) principal component analysis (Barreiro Lozano, 1991; Carral et al., 1995a, 1995b; Chester & Voutsinou, 1981; Hakanson, 1984; Rubio et al., 2000b; Summers et al., 1996).

The selection of the BL will conditionate the geochemical interpretation of a particular area (Rubio et al., 2000a). The use of local or regional BLs is recommended to take into account local geochemical variability. Even in a local study, BL should be selected considering the watershed composition. It is important to highlight that sediments can be enriched naturally in a particular trace element. For example, Álvarez-Iglesias et al. (2006) showed that sediments in inner Ría de Vigo from a schist-gneiss watershed were enriched in metals such as Fe, Cr and Ni when compared with sediments from a granitic watershed (Fig. 7). Circles in Fig. 7 correspond to a subtidal core of inner ría with relatively high Fe/Al ratios, pointing to a schist-gneiss source, in comparison to another subtidal core samples from inner ría. This can also be deduced for middle and outer Ría de Vigo sediments, with high Fe/Al ratios too. Grain-size and postdepositional physico-chemical processes can also affect to remobilization of some redox-sensitive elements. For example, it can be observed in Fig. 7 that some of the intertidal samples (light squares) showed higher Fe/Al ratios than others from the same area. This can be interpreted as Fe enrichments caused by diagenetic processes. Furthermore, it can be observed that intertidal samples of inner ría (Fig. 7, squares) showed higher coarse-grained particles content and lower concentrations of Fe and Ti than subtidal samples of the ría. Then, it is recommended the use of BLs with similar characteristics (grain size, composition, origin) to those of the sediments of the area under study to avoid misinterpretation of the data. It is also recommended the use of BLs obtained by similar analytical procedures than those applied on the target samples. This will avoid the possible differences in metal concentrations related only to the efficiency of the followed protocol (Rubio et al., 2000a). For example, Rubio et al. (2000b) established a BL for Cr by XRF higher than that previously established for the same study area by triacid digestion (Rubio et al., 2000a) (Table 4).

Scale	Al	Ca	Fe	K	Mg	P	S	Si	Ti
Global[1,a]	8.00	2.21	4.72	2.66	1.50	0.07	0.24	7.30	0.46
Global[1,b]	8.40	2.90	6.50	2.50	2.10	0.15	0.13	25.00	0.46
Global[1,c]	7.20	0.51	1.42	4.2	0.16	0.06	0.03	34.7	0.12
Global[2,d]	8.89	1.64	4.83	2.82	1.57	-	-	27.11	0.43
Global[3,e]	8.23	4.15	5.63	2.09	2.33	0.105	0.026	28.15	0.57
Global[3,f]	7.70	1.58	2.70	3.34	0.16	0.070	0.027	32.3	0.23
Galicia[4]	-	-	2.69	-	-	-	-	-	-
Galicia[5]	-	-	2.6	-	-	-	-	-	-
Galicia[6,g]	-	-	2.9	-	-	-	-	-	-
Galicia[6,h]	-	-	3.3	-	-	-	-	-	-
RV[7]	6.48	-	3.51	-	-	-	-	-	0.34
RV[8]	6.48	-	3.51	-	-	-	-	-	0.34
SSB[9]	9.82	3.16	3.53	2.59	1.05	0.10	1.06	21.49	0.36
SSB[10,i]	9.91	3.22	3.36	2.56	1.05	0.10	0.94	21.45	0.36
SSB[10,j]	9.36	2.81	4.44	2.75	1.05	0.08	1.69	21.70	0.37

[1] Turekian & Wedepohl (1961); [2]Wedepohl (1971, 1991); [3]Taylor (1964); [4]Barreiro Lozano (1991); [5]Carral et al. (1995a); [6]Carral et al. (1995b); [7]Rubio et al. (2000a); [8]Rubio et al. (2000b); [9]Álvarez-Iglesias et al. (2006); [10]This study. [a]shales, [b]deep-sea clay sediments, [c]low Calcium granites, [d]average shales, [e]Earth crust, [f]granite average, [g]estuarine sediments from granitic watersheds, [h]estuarine sediments from schist-gneiss watersheds; [i]ría sediments with a granitic source; [j]ría sediments with a schist-gneiss source. RV: Ría de Vigo, SSB: San Simón Bay (inner Ría de Vigo).

Table 3. Global, regional and local background levels for major elements (%).

Scale*	As	Cr	Cu	Mn	Ni	Pb	Rb	V	Zn
Global[1,a]	13	90	45	850	68	20	140	130	95
Global[1,b]	13	90	250	6700	225	80	110	120	165
Global[1,c]	1.5	4.1	10	390	4.5	19	170	44	39
Global[2,d]	-	90	-	-	68	-	-	130	95
Global[3,e]	1.8	100	55	950	75	12.5	90	135	70
Global[3,f]	1.5	4	10	400	0.5	20	150	20	40
Galicia[4]	-	43	25	225	30	25	-	-	100
Galicia[5]	-	32	28	275	32	53	-	-	122
Galicia[6,g]	-	30	20	248	31	78	-	-	136
Galicia[6,h]	-	54	35	395	38	50	-	-	120
RV[7]	-	34	29	244	30	51	-	-	105
RV[8]	-	55	20	244	30	25	-	-	105
SSB[9]	16	65	21	216	33	51	198	94	110
SSB[10,i]	14	64	19	214	33	54	180	93	111
SSB[10,j]	19	70	30	229	35	34	298	99	104

*Uppercase notation similar to that of Table 3.

Table 4. Global, regional and local background levels for minor and trace elements (µg g[-1]).

The first step for evaluating the degree of pollution of inner Ría de Vigo sediments would be to make the scatterplots of Al-Fe, Al-K, Al-S, Al-Si , Al-Ti and Al-Rb (Fig. 7) in order to decide which background value have to be selected for obtaining reliable data. Those samples out of the general trend in the scatter plots will be identified as enriched samples. In the samples under study, it is neccesary to consider two different background values: one BL for the samples from middle and outer ría together with the samples of one subtidal core (SS4), and another BL for the rest of the subtidal core samples and the intertidal core samples from the inner ría.

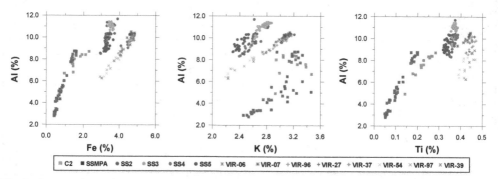

Fig. 7. Scatterplots of Al vs Fe, Al vs K and Al vs Ti in core samples (N=210) from different sectors of the Ría de Vigo (sample location in Fig. 1). Metal contents from XRF analyses. Note the marked differences between intertidal samples (with a higher coarse-grained particles content) and subtidal samples, and the striking differences within subtidal samples, where two populations can be observed, one, with higher Fe/Al ratios (those samples with a schist-gneiss source), the other, with lower Fe/Al ratios (most of the inner ría subtidal cores).

5.2 Degree of pollution

The indexes that are usually applied to evaluate the presence of metal pollution in sediments are the Contamination or Concentration Factor (CF) and the Enrichment Factor (EF) (Barreiro Lozano, 1991; Álvarez-Iglesias et al., 2003; Rubio et al., 2000a).

The CF (Hakanson, 1980) is based on the calculation of the relationship between the concentration of the potential pollutant element measured in the sample $[M]_i$ and the background level for that element $[M]_{BL}$. $FC=[M]_i/[M]_{BL}$. CF value would be close o 1 when there is no metallic enrichment. The degree of contamination can be classified as null or low (CF<1), moderate (CF: 1-3), high (CF: 3-6) or very high (CF>6).

The EF (Zoller et al., 1974) is based on the calculation of the relationship between the concentration of the potential pollutant element and a normalizer element measured in the sample ($[M]_i$ and $[N_e]_i$, respectively), divided by the same relationship considering the previously selected BL ($[M]_{BL}$ and $[N_e]_{BL}$, respectively): $FE=([M]_i/[N_e]_i)/([M]_{BL}/[N_e]_{BL})$. An EF=1 indicates non-enrichment in comparison with the BL. Nevertheless, an EF lower/higher than 1 points to an impoverishment/enrichment, respectively, of the studied trace element. The degree of pollution can be classified in a similar way to that of CF.

Both CF and EF has been calculated for the different cores studied using the new BLs established (Fig. 8). The dispersion among cores considering the deepest samples is higher for CF than for EF depth-profiles, this is explained by the non-consideration of grain-size variability when calculating CF. Furthermore, the high grain-size variability of the intertidal cores, where coarse-grained sediments are very abundant at certain levels, explains the impoverishment observed for the considered trace metals at their bottom. Both CF and EF indexes show a moderate enrichment in Cu and Zn, in general, for the first 0.10-0.20 m of the sedimentary record, in general, for all the cores. Nevertheless, Pb pollution has been detected in the first 0.10-0.80 m, its degree varying from moderate to very high depending on core depth and position. Pollution was classified as high in the first decimeter of inner and middle subtidal cores, whereas the highest EF (up to 40-45) has been detected in the first 0.15-0.20 m of the intertidal core samples. These moderate or high contents of trace metals were explained as anthropogenic inputs, in particular for Cu and Zn from fertilizers and pesticides (Álvarez-Iglesias et al., 2006, 2007) and for Pb both from direct and indirect sources. The main Pb input since the 1970s to the inner ría has been a ceramic factory located at the head of the Ría but previously and nowadays diffuse sources, mainly coal and petrol combustion, have contributed and contribute to Pb inputs to the Ría de Vigo (Álvarez-Iglesias et al., 2003, 2007).

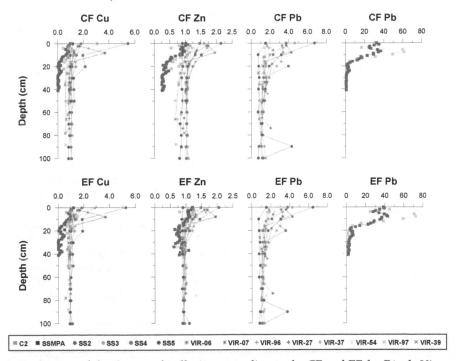

Fig. 8. Evaluation of the degree of pollution according to the CF and EF for Ría de Vigo core samples (sample location in Fig. 1). Note that for the particular case of the subtidal samples of the inner ría it has been only considered those from the first sampled meter in order to allow for a better comparison between cores. CF-Pb and EF-Pb for intertidal cores are showed in a separated graph because of their high pollution factors.

6. Early diagenesis of metals in marine sediments

Early diagenesis processes include all the post-depositional changes that happen in sediments in a short geological time-scale. After sediment deposition, its constituents can suffer chemical modifications, implying a new equilibrium between solid and dissolved species. Then, metal speciation can be modified, and thus, metal mobility and bioavailability too (Shaw et al., 1990). Organic matter degradation fuels early diagenesis processes. Their mineralization is mainly bacterially mediated following an ideal sequence of alternative electron acceptors (Froelich et al., 1979), where the most thermodynamically favoured reaction happens until exhaustion of the corresponding reactant, and then, the next favourable reaction starts. In this way, a diagenetic zonation can be described for the Ría de Vigo sediments.

6.1 Diagenetic zonation

Diagenetic zonation in marine sediments depends on the degree of oxygenation of the water column (Wignall, 1994). According to Berner (1981) different sedimentary environments can be differentiated considering redox potential, electron aceptors and typical mineral associations: oxic, suboxic, anoxic-sulphidic and anoxic-non sulphidic. These four diagenetic zones can be observed in sediments below oxic bottom waters and can be inferred by metal speciation procedures. Considering that sulphate reduction is a significant process in marine sediments, the protocols where the sulphide-bound fraction is independently obtained will be preferred (i.e., H&M's method). One index that is usually used is the degree of pyritization (DOP; Berner, 1970) where the relationship between the pyrite-bound and the reactive fractions is calculated. This index has recently been regarded as an indicator of interstitial water status in coastal sediments (Roychoudhury et al., 2003), with the limits for oxic with DOPs lesser than 42%, suboxic between 42 and 55%, anoxic between 55 and 75% and euxinic higher than 75% (León et al., 2004; Raiswell et al., 1988).

The oxic zone is characterized by the presence of oxygen, which is the first electron acceptor in the organic matter degradation process. Vertical diffusion of dissolved oxygen from the adjacent water column maintains oxygenated the interstitial water of freshly deposited sediments. Nevertheless, oxygen is rapidly consumed with depth because of the high metabolic rates of aerobic bacteria, avoiding vertical diffusion. Organic matter oxidation generates carbon dioxide, nitrate and phosphate (Froelich et al., 1979). CO_2 generation does not usually cause a pH diminution in the interstitial waters, but pH slightly diminishes because of an extensive oxidation of those sulphides migrating toward surface which also causes a carbonate subsaturation. Sediments contains oxides and oxyhydroxides of Fe (hematite, goethite) and Mn, and usually, a low content of organic matter, which is mostly degraded. The extension of the oxic zone is reduced, typically 0.10 m, but controlled by the rate of oxygen comsuption which depends on organic matter abundance and sedimentation rate (Wignall, 1994). Bioturbation by organisms in this zone is significant. In inner Ría de Vigo sediments the oxic zone is extended, as a minimum, along the first 0.25 m of the sedimentary record in intertidal coarse-grained sediments (sediments with about 90% of sand and gravel; Álvarez-Iglesias & Rubio, 2009) but it is restricted to the first 0.06 m in sediment cores with a higher fine-grained particles content (around 37%, core C2). In subtidal inner and middle ría sediments oxic conditions hold only in the top few millimetres (Rubio et al., 2010; Álvarez-Iglesias & Rubio, 2012), whereas the oxic zone is restricted to the top decimetre of outer ría sediments (Rubio et al., 2010).

The suboxic zone is characterized by a very low concentrations of oxygen. Here nitrate reduction, Mn reduction and Fe reduction take place (Froelich et al., 1979). Once interstitial water O_2 is consumed, a narrow denitrification zone appears where nitrate is the oxidant. This process accounts for a small percent of the degradation of organic matter (Canfield & Raiswell, 1991). It is also observed an extensive oxidation of those sulphides migrating towards surface, generating elemental Sulphur and sulphate. Then, interstitial waters in the suboxic zone are slightly acidic, and dissolution of carbonated shells takes place. Sediments show glauconite and other Fe^{+2}-Fe^{+3} silicates, Mn carbonates (rhodocrosite), Fe carbonates (siderite), Fe phosphates (vivianite), non-sulphidic minerals and, usually, a low content of organic matter. In inner Ría de Vigo the suboxic zone is extended approximately from 0.06 to 0.22 m in intertidal sediment cores with a high fine-grained particles content (Álvarez-Iglesias & Rubio, 2009). In subtidal inner and middle ría sediments, the suboxic conditions extend down to 0.10-0.20 m, and in outer ría sediments, down to 0.40 m, excepting those areas of strong turbulent currents (core SS5), where suboxic conditions extend, as a minimum, down to a few meters (Rubio et al., 2010; Álvarez-Iglesias & Rubio, 2012).

Buried Mn oxides below surficial oxic layer are reductively dissolved in the suboxic layer and Mn^{+2} are released to the interstitial waters. These ions diffuse towards surface because of a concentration gradient, reprecipiting as a "second generation", "third generation", etc., of oxides in the bottom of the oxic layer. Then, it can be generated sedimentary layers highly enriched in Mn over the Mn-redox limit as a result of Mn oxidative reprecipitation in the oxic layer (called "Mn trap"). Similarly, Fe^{2+} diffusion towards surface will generate Fe oxyhydroxides reprecipitation. Then, Fe and Mn profiles can be highly affected by redox changes. This has clearly been observed in intertidal sand flat sediments of inner Ría de Vigo, where subsurficial maxima were detected for both elements related to Fe and Mn ions migration and reprecipitation (Álvarez-Iglesias et al., 2003). It has also been detected in middle and outer ría sediments, where it has been observed Mn reducible (BCR method) maxima in the metal content profiles (Rubio et al., 2010). Associated to those maxima it has also been observed Cu, Pb and Zn maxima in the reducible fraction of Ría de Vigo sediments, indicating trace metal coprecipitation with the authigenic oxyhydroxides.

The anoxic-sulphidic zone or sulphate-reduction zone (SRZ) is characterized by metal sulphide generation (Huerta-Díaz & Morse, 1992). Strict anaerobic bacteria run sulphate-reduction, which is a dominant process in marine sediments, in particular, this process accounts for about 50 % of organic matter degradation in coastal sediments (Jørgensen, 1982). When organic matter is degraded by sulphate reduction weak acids are generated (HCO_3^-, HS^-, HPO_4^{2-}), then carbonate subsaturation (coming from the previous zone) is maintained. Most of the generated sulphide ions are oxidized in the redox upper boundary but a significant proportion reacts with iron to form, in the last step, pyrite. The main Fe sources are detrital Fe minerals, in particular fine-grained Fe oxides. Sulphide ions can also react with organic matter and form organosulphur compounds, which represents usually a small fraction in coastal sediments (Berner, 1970, 1981; Fig. 9). While sulphide is removed from dissolution by sulphide generation, interstitial waters pH increase and then, carbonate saturation happens. Then, typical minerals in the SRZ are pyrite, marcasite, rhodocrosite, and alabandite. Different metal ions can coprecipitate with authigenic pyrite or form other sulphides, then, their bioavailability will be limited while environmental reducing conditions maintain. In anoxic sediments from the inner sector of Ría de Vigo it has been observed that ~28% of Cu, ~34% of Pb and ~1.5% of Zn is retained in the pyrite-bound fraction in intertidal sediments, whereas ~82% of Cu, ~18% of Pb and ~5.1% of Zn in the same fraction of subtidal sediments (Álvarez-Iglesias & Rubio, 2008, 2009). In middle and outer ría sediments it has been observed, in general, an increase with depth in the contents of Cu and Zn in the oxidizable fraction (BCR method) that would probably be related to sulphide generation, taking into account that the organic fraction accounted for ~1.2% of Cu and ~5.2% of Zn in anoxic sediments and ~2.3% of Cu and ~5.1% of Zn in suboxic sediments from the subtidal inner ría sector.

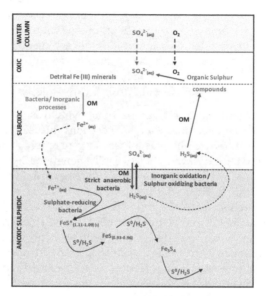

Fig. 9. Schematic representation of pyrite formation in marine sediments (modified from Berner (1970), Rickard et al. (1995) and Schoonen & Barnes (1991). OM= organic matter.

At low temperatures pyrite formation can be conceptually reduced to two steps: a) precipitation of an iron monosulphide precursor (FeS*), and b) pyrite formation by S addition (Berner, 1970; Rickard et al., 1995; Schoonen & Barnes, 1991) (Fig. 9). Nevertheless, direct pyrite precipitation has been described in saltmarsh sediments without intermediate precursors (Giblin & Howard, 1984). The most common pyrite morphologies described in Galician rías sediments, are euhedral and single framboids or polyframboids (Rubio et al., 2001; Álvarez-Iglesias & Rubio, 2012).

In the SRZ it could be observed Mn peaks. Alcalinity genesis in this zone is avoided when organism galleries penetrate actively and regularly. This causes sulphide oxidation. Carbonate dissolution may happen too (Aller, 1982; Canfield & Raiswell, 1991). The limit between the suboxic zone and the SRZ is called redoxcline. It represents a barrier to element transport by diffusion because of the presence of amorphous Fe-Mn oxyhydroxides. It is a dynamic limit, changing its position temporally and spatially. It can be detected de visu in intertidal sediments, such as those of inner Ría de Vigo, by the colour change between the brown-reddish oxidized surficial sediments –which reflects Fe oxides precipitation- and the dark grey-black reduced sediments-which reflects pyrite precipitation. The anoxic zone develops from 0.14 to 0.18 m in those intertidal sediment cores from the inner ría with a high fine-grained particles content (core C2; Álvarez-Iglesias & Rubio, 2009). In subtidal sediments from inner and middle ría the anoxic zone is extended down the first one-two decimetres of the sedimentary record (with the exception of turbulent areas), whereas in the outer zone, down the first 0.40 m (Rubio et al., 2010; Álvarez-Iglesias & Rubio, 2012). Then, the diagenetic zones all became thicker, as well as lying deeper, as the mouth of the ría was approached.

Finally, the anoxic-non sulphidic zone or methanogenic zone (MZ) is characterized by methane generation by bacterial fermentation (Froelich et al., 1979) which causes a slight pH diminution. Typical minerals in the MZ are siderite, vivianite, rhodocrosite and previously formed sulphides. The MZ only developes if high organic matter quantities have not been degraded in the SRZ. Then, its development is favoured in areas with a high organic carbon flux to the sediments and/or high sedimentation rates, where up to 10% of the sedimentary organic matter is consumed by methanogenesis. When methane solubility is exceeded, gas bubbles accumulate and methane gas will diffuse upwards. In subtidal sediments of the Ría de Vigo it has been described several shallow gas fields, where biogenic methane was detected (García-Gil et al., 2002; Iglesias & García-Gil, 2007) and gas bubbles have been observed in the sediments (Álvarez-Iglesias et al., 2006; García-Gil et al., 2002). The top of the gas fields was located at sediment depths ranging between 11 and 0 m, shallowing towards the ría head (García-Gil et al., 2002; Iglesias & García-Gil, 2007).

7. Conclusion

The study of the sedimentary record allows establishing metal pollution history spatially and temporally for a particular area and needs to be multidisciplinary. First, hydrological, hydrographical, geological and anthropological information (industrial and rural development) on the study area needs to be compiled. Second, adequate analytical techniques, reference materials and background levels have to be selected. Background levels and metal concentrations in target samples should be analyzed by the same analytical protocols.

It has to be considered the relationships between metal concentrations, mineralogy, grain-size and metal sources. Sediments from Ría de Vigo showed that lithogenic elements such as Al, V or Ni were more abundant in fine-grained sediments and elements such as Si, in coarse grained sediments, showing the typical association of trace metals to organic-rich fine-grained sediments. Furthermore, biogenic inputs caused a dilution of detrital metal inputs. For correction of this grain-size effect, the use of normalization techniques is highly recommended.

It is also recommended the use of local or regional background levels to take into account the geological variability of the study area. It is useful for its selection to analyze deep sediments with no-human influence and to use metal-Al scatter-plots to identify outliers (enriched samples). This procedure has shown that in the Ría de Vigo sediments two clear sediment populations can be differentiated influenced by the watersheds composition (metallic natural source) and two different BLs have been considered, depending on the population.

Metal pollution can be identified by the use of pollution indexes, such as the Contamination Factor and the Enrichment Factor. These indexes showed that sediments of Ría de Vigo are moderatelly polluted by Cu and Zn in the first one-two decimeters of the sedimentary record, whereas moderatelly to very highly polluted by Pb, specially in the inner ría sediments.

Metal concentration alone it is not enough to evaluate metal availability and mobility, then, metal speciation protocols have to be considered. The protocols usually followed are those non-selective. In Ría de Vigo sediments, in general, a significant percentage of Cu, Pb and Zn is hosted in the residual fraction, whereas a small content is retained in the interchangeable and bound to carbonates fraction. Metal fraction content bound to Fe-Mn oxyhydroxides is higher in intertidal than subtidal inner ría sediments. Nevertheless, the reactive fraction content contribution is higher in middle and outer ría sediments. Concentrations of Cu, Pb and Zn are very low in the labile fraction of middle and outer ría sediments, and also, the Zn contents in the interchangeable and bound to carbonates fractions. Nevertheless, Pb concentrations in these last two fractions-that are highly bioavailable- in intertidal sediments oversize the background levels established for total concentrations.

The modification of metal speciation patterns by diagenetical processes has been showed for Ría de Vigo sediments, where Mn enrichments were detected, clearly in sandy intertidal sediments. Different sedimentary environments –oxic, suboxic, anoxic-non sulphidic- and their extension has been detected in the studied ría sediments. The boundaries between these diagenetic zones became deeper towards the ría mouth.

8. Acknowledgment

This work was supported by the Spanish Ministry of Science and Technology through projects CTM2007-61227/MAR, GCL2010-16688 and IPT-310000-2010-17, by the IUGS-UNESCO through project IGCP-526 and by the Xunta de Galicia through projects 09MMA012312PR and 10MMA312022PR.

9. References

Ackermann, F. (1980). A procedure for correcting the grain size effect in heavy metal analyses of estuarine and coastal sediments. *Environmental Technology Letters*, Vol. 1, pp. 518-527.

Aller, R.C. (1982). Carbonate dissolution in nearshore terrigenous muds—the role of physical and biological reworking: *Journal of Geology*, Vol. 90, pp. 79-95.

Álvarez-Iglesias, P.; Quintana, B., Rubio, B. & Pérez-Arlucea, M. (2007). Sedimentation rates and trace metal input history in intertidal sediments derived from ^{210}Pb and ^{137}Cs chronology. *Journal of Environmental Radioactivity*, Vol. 98, pp. 229-250.

Álvarez-Iglesias, P. & Rubio, B. (2008). The degree of trace metal pyritization in subtidal sediments of a mariculture area: application to the assessment of toxic risk. *Marine Pollution Bulletin*, Vol. 56, pp. 973–983.

Álvarez-Iglesias, P. & Rubio, B. (2009). Redox status and heavy metal risk in intertidal sediments in NW Spain as inferred from the degrees of pyritization of iron and trace elements. *Marine Pollution Bulletin*, Vol. 58, pp. 542-551.

Álvarez-Iglesias, P. & Rubio, B. (2012). Early diagenesis of organic-matter-rich sediments in a ría environment: Organic matter sources, pyrites morphology and limitation of pyritization at depth. *Estuarine, Coastal and Shelf Science*, Vol. 100, pp. 113-123.

Álvarez-Iglesias, P.; Rubio, B. & Pérez-Arlucea, M. (2006). Reliability of subtidal sediments as "geochemical recorders" of pollution input: San Simón Bay (Ría de Vigo, NW Spain). *Estuarine, Coastal and Shelf Science*, Vol. 70, pp. 507-521.

Álvarez-Iglesias, P.; Rubio, B. & Vilas, F. (2003). Pollution in intertidal sediments of San Simón Bay (Inner Ría de Vigo, NW of Spain): total heavy metal concentrations and speciation. *Marine Pollution Bulletin*, Vol. 46, pp. 491-506.

Álvarez-Salgado, X. A.; Gago, J.; Míguez, B. M.; Gilcoto, M. & Pérez, F. F. (2000). Surface waters of the NW Iberian Margin: upwelling on the shelf versus outwelling of upwelled waters from the Rías Baixas. *Estuarine, Coastal and Shelf Science*, Vol. 51, pp. 821-837.

Barreiro Lozano, R. (1991): *Estudio de metales pesados en medio y organismos de un ecosistema de ría (Pontedeume, A Coruña)*. Ph.D. Thesis, Universidad de Santiago de Compostela, 227 p.

Belzunce-Segarra, M.J.; Bacon, J.R.; Prego, R. & Wilson, M.J. (1997). Chemical forms of heavy metals in surface sediments of the San Simon inlet, Ria de Vigo, Galicia. *Journal of Environmental Science and Health*, Vol. A32, pp. 1271-1292.

Berner, R.A. (1970). Sedimentary pyrite formation. *American Journal of Science*, Vol. 268, pp. 1-23.

Berner, R.A. (1981). A new geochemical classification of sedimentary environments. *Journal of Sedimentary Petrology*, Vol. 51, No.2, pp. 359-365.

Borovec, Z. (1996). Evaluation of the concentrations of trace elements in stream sediments by factor and cluster analysis and the sequential extraction procedure. *The Science of the Total Environment*, Vol. 177, pp. 237-250.

Bryan, G.W. & Langston, W.J. (1992). Bioavailability, accumulation and effects of heavy metals in sediments with special reference to United Kingdom Estuaries: a review. *Environmental Pollution*, Vol. 76, pp. 89-131.

Boyle, J.F. (2000). Rapid elemental analysis of sediment samples by isotope source XRF. *Journal of Paleolimnology*, Vol. 23, pp. 213-221.

Campbell, P.G.C.; Lewis, A.G.; Chapman, P.M.; Crowder, A.A.; Fletcher, W.K.; Imber, B.; Luoma, S.N.; Stokes, P.M. & Winfrey, M. (1988). *Biologically available metals in sediments*. NRCC, Division of Chemistry Publication, Otawa, 296 pp.

Canfield, D.E. & Raiswell, R. (1991). Pyrite formation and fossil preservation, *Taphonomy: Releasing the Data Locked in the Fossil Record. Topics in Geobiology*. P.A. Allison & D.E.G. Briggs (Eds.), Vol. 9, 337–387.

Carral, E.; Puente, X.; Villares, E. & Carballeira, A. (1995a). Background heavy metal levels in estuarine sediments and organism in Galicia (Northwest Spain) as determined by modal analysis. *Science of the Total Environment*, Vol. 172, pp. 175-188.

Carral, E.; Villares, R.; Puente, X. & Carballeira, A. (1995b). Influence of watershed lithology on heavy metal levels in estuarine sediments and organisms in Galicia (North-west Spain). *Marine Pollution Bulletin*, Vol. 30, No.9, pp. 604-608.

Chester, R. & Voutsinou, F. G. (1981). The initial assessment of trace metal pollution in coastal sediments. *Marine Pollution Bulletin*, Vol. 12, pp. 84-91.

Dassenakis, M.I.; Kloukiniotou, M.A. & Pavlidou, A.S. (1996). The influence of long existing pollution on trace metal levels in a small tidal Mediterranean Bay. *Marine Pollution Bulletin*, Vol. 32, pp. 275-282.

Duffus, J.H. (2002). "Heavy metals" – A meaningless term? *Pure Applied Chemistry*, Vol. 74, pp. 793-807.

Fichet, D.; Radenac, G. & Miramand, P. (1998). Experimental studies of impacts of harbour sediments resuspension to marine invertebrates larvae: bioavailability of Cd, Cu, Pb and Zn and toxicity. *Marine Pollution Bulletin*, Vol. 36, pp. 509-518.

Flyhammar, P. (1998). Use of sequential extraction on anaerobically degraded municipal solid waste. *The Science of the Total Environment*, Vol. 212, pp. 203-215.

Förstner, U. (1989). *Contaminated Sediments. Lecture notes on Earth Science 21.* Springer-Verlag, London, 157 pp.

Freire, J. & García-Allut, A. (2000). Socioeconomic and biological causes of management failures in European artisanal fisheries: the case of Galicia (NW Spain). *Marine Policy*, Vol. 24, pp. 375-384.

Froelich, P.N.; Klinkhammer, G.P.; Bender, M.L.; Luedtke, N.A.; Heath, G.R.; Cullen, D.; Dauphin P.; Hammond, D. & Hartman, B. (1979). Early oxidation of organic matter in pelagic sediments of the eastern equatorial Atlantic: suboxic diagenesis. *Geochimica et Cosmochimica Acta*, Vol. 43, pp. 1075-1090.

García-Gil, S.; Vilas, F. & García-García, A. (2002). Shallow gas features in incised-valley fills (Ria de Vigo, NW Spain): a case study. *Continental Shelf Research*, Vol. 22, pp. 2303-2315.

Giblin, A.E. & Howarth, R.W. (1984). Porewater evidence for a dynamic sedimentary iron cycle in salt marshes. *Limnology and Oceanography*, Vol. 29, pp. 47-63.

Gómez Ariza, J.L.; Giráldez, I.; Sánchez-Rodas, D. & Morales, E. (2000). Comparison of the feasibility of three extraction procedures for trace metal partitioning in sediments from South-West Spain. *The Science of the Total Environment*, Vol. 246, pp. 271-283.

González-Pérez, J.M. & Pérez González, A. (2003). Demographic dinamycs and urban planning in Vigo since 1960. The impact of industrialization. *Anales de Geografía de la Universidad Complutense*, Vol. 23, pp. 163-185.

Guy, R.D.; Chakrabarti, C.L. & McBain, D.C. (1978). An evaluation of extraction techniques for the fractionation of Copper and Lead in model sediment systems. *Water Research*, Vol. 12, pp. 21-24.

Hakanson, L. (1980). An ecological risk index for aquatic pollution control. A sedimentological approach. *Water Research*, Vol. 14, pp. 975-1001.

Hakanson, L. (1984). Metals in fish and sediments from the River Kolbäcksan water system, Sweden. *Archiv für Hydrobiologie*, Vol. 101, No.3, pp. 373-400.

Horowitz, A.J. & Elrick, K.A. (1987). The relation of stream sediment surface area, grain size, and composition of trace element chemistry. *Applied Geochemistry*, Vol. 2, pp. 437-451.

Howard, J.L & Shu, J. (1996). Sequential extraction analysis of heavy metals using a chelating agent (NTA) to counteract resorption. *Environmental Pollution*, Vol. 91, No. 1, pp. 89-96.

Huerta-Díaz, M.A. & Morse, J.W. (1992). Pyritisation of trace metals in anoxic marine sediments. *Geochimica et Cosmochimica Acta*, Vol. 56, pp. 2681-2702.

Huerta-Díaz, M.A. & Morse, J. (1990). A quantitative method for determination of trace metal concentrations in sedimentary pyrite. *Marine Chemistry*, Vol. 29, pp. 119-144.

IGME (1981). *Mapa Geológico de España*, 1:50000, Servicio de Publicaciones Ministerio de Industria y Energía.

Imperato, M.; Adamo, P.; Naimo, D.; Arienzo, M.; Stanzione, D. & Violante, P. (2003). Spatial distribution of heavy metals in urban soils of Naples city (Italy). *Environmental Pollution*, Vol. 124, pp. 247-256.

Iglesias, J. & García-Gil. S., 2007. High-resolution mapping of shallow gas accumulations and gas seeps in San Simón Bay (Ría de Vigo, NW Spain). *Geo-Marine Letters*, Vol. 27, pp. 103-114.

Izquierdo, C.; Usero, J. & Gracia, I. (1997). Speciation of heavy metals in sediments from salt marshes on the Southern Atlantic Coast of Spain. *Marine Pollution Bulletin*, Vol. 34, No. 2, pp. 123-128.

Jørgensen, B.B. (1982). Mineralization of organic matter in the sea bed- the role of sulphate reduction. *Nature*, Vol. 296, pp. 643-645.

Kersten, M. & Förstner, U. (1987). Effect of sample pretreatment on the reliability of solid speciation data of heavy metals -implications sesfor the study of early diagenetic processes. *Marine Chemistry*, Vol. 22, pp. 99-312.

León, I.; Méndez, G. & Rubio, B. (2004). Geochemical phases of Fe and degree of pyritization in sediments from Ría de Pontevedra (NW Spain): Implications of mussel raft culture. *Ciencias Marinas*, Vol. 30, pp. 585-602.

Loring, D.H. (1981). Potential bioavailability of trace metals in eastern Canadian estuarine sediments. *Rapports et Proces-Verbaux des Reunions, Conseil International pour l'Exploration de la Mer*, Vol. 181, pp. 93-101.

Loring, D.H. (1990). Lithium-a new approach for the granulometric normalization of trace metal data. *Marine Chemistry*, Vol. 29, pp. 155-168.

Luoma S.N. & Bryan, G.W. (1981). A statistical assessment of the form of trace metals in oxidized estuarine sediment employing chemical extractants. *The Science of the Total Environment*, Vol. 17, pp. 165-196.

Luoma, S.N. (1986). A comparison of two methods for determining copper partitioning in oxidized sediments. *Marine Chemistry*, Vol. 20, pp. 45-49.

Mahan, K. I.; Foderaro, T.A.; Garza, T.L.; Martínez, R.M.; Maroney, G.A.; Trivisonno, M.R. & Willging, E.M. (1987). Microwave digestion techniques in the sequential extraction of Calcium, Iron, Chromium, Manganese, Lead and Zinc in sediments. *Analytical Chemistry*, Vol. 59, pp. 938-945.

Moore, H.W. (1991). *Inorganic Contaminants of Surface Water. Research and Monitoring priorities*. Springer-Verlag, New York, 334 pp.

Ngiam, L.-S. & Lim, P.-E. (2001). Speciation patterns of heavy metals in tropical estuarine anoxic and oxidized sediments by different sequential extraction schemes. *The Science of the Total Environment*, Vol. 275, pp. 53-61.

Pérez-Arlucea, M.; Méndez, G.; Clemente, F.; Nombela, M.; Rubio, B. & Filgueira, M. (2005). Hydrology, sediment yield, erosion and sedimentation rates in the estuarine environment of the Ría de Vigo, Galicia, Spain. *Journal of Marine Systems*, Vol. 54, pp. 206-226.

Prego, R. & Cobelo-Garcia, A. (2003). Twentieth century overview of heavy metals in the Galician Rias (NW Iberian Peninsula). *Environmental Pollution*, Vol. 121, pp. 425–452.

Prego, R., Otxotorena, U. & Cobelo-García, A. (2006) Presence of Cr, Cu, Fe and Pb in sediments underlying mussel-culture rafts (Arosa and Vigo rias, NW Spain). Are they metal-contaminated areas? *Ciencias Marinas*, Vol. 32, No.2B, pp. 339-349.

Quevauviller, Ph.; Rauret, G.; López-Sánchez, J-F.; Rubio, R.; Ure, A. & Muntau, H. (1997). Certification of trace metal extractable contents in a sediment reference material (CMR 601) following a three-step sequential extraction procedure. *The Science of the Total Environment*, Vol. 205, pp. 223-234.

Raiswell, R.; Buckley, F.; Berner, R. & Anderson, T. (1988). Degree of pyritization of iron as a paleoenvironmental indicator of bottom-water oxygenation. *Journal of Sedimentary Petrology*, Vol. 58, pp. 812-819.

Rapin, F. & Förstner, U. (1983). Sequential leaching techniques for particulate metal speciation: the selectivity of various extractants, In: *Proceedings of the 4th Conference on Heavy Metals and the Environment* (Heidelberg, Germany), 1074-1077.

Rendell, P.S.; Bately, G.E. & Cameron, A.J. (1980). Adsorption as a control of metal concentrations in sediment extracts. *Environmental Technology Letters*, Vol. 14, pp. 314-318.

Rey, D.; Mohamed, K.J.; Bernabeu, A.; Rubio, B. & Vilas, F. (2005). Early diagenesis of magnetic minerals in marine transitional environments: geochemical signatures of hydrodynamic forcing. *Marine Geology*, Vol. 215, pp. 215-236.

Rickard, D.; Schoonen, M.A.A. & Luther III, G.W. (1995). Chemistry of iron sulfides in sedimentary environments, In: *Geochemical Transformations of Sedimentary Sulfur*, M.A. Vairavamurthy & M.A.A. Schoonen (Eds.), Vol. 612, 168-193. American Chemical Society Symposium Series.

Roychoudhury, A.N.; Kostka, J.E. & Van Capellen, P. (2003). Pyritization: a palaeoenvironmental and redox proxy reevaluated. *Estuarine, Coastal and Shelf Science*, Vol. 57, pp. 1183-1193.

Rubio, B.; Álvarez-Iglesias, P. & Vilas, F. (2010). Diagenesis and anthropogenesis of metals in the recent Holocene sedimentary record of the Ría de Vigo (NW Spain). *Marine Pollution Bulletin*, Vol. 60, pp. 1122-1129.

Rubio, B.; Nombela M.A. & Vilas F. (2000a). Geochemistry of major and trace elements in sediments of the Ría de Vigo (NW Spain): An assessment of metal pollution. *Marine Pollution Bulletin*, Vol. 40, pp. 968-980.

Rubio, B.; Nombela M.A. & Vilas F. (2000b): La contaminación por metales pesados en las Rías Baixas gallegas: nuevos valores de fondo para la Ría de Vigo (NO de España). *Journal of Iberian Geology*, Vol. 26, pp. 121-149.

Rubio, B.; Pye, K.; Rae, J. & Rey, D. (2001). Sedimentological characteristics, heavy metal distribution and magnetic properties in subtidal sediments, Ría de Pontevedra, NW Spain. *Sedimentology*, Vol. 48 No.6, pp. 1277-1296.

Rubio, B.; Rey, D.; Bernabeu, A.; Vilas, F. & Rodríguez-Germade, I. (2011). Nuevas técnicas de obtención de datos geoquímicos de alta resolución en testigos sedimentarios, el XRF core scanner, In: *Métodos y técnicas en investigación marina*. Editorial TECNOS, Madrid (Spain), 383-393.

Salomons, W. & Förstner, U. (1984). *Metals in the hydrocycle*. Springer-Verlag, Berlin, 349 p.

Salomons, W.; Kerdiijk, H.; Van Pagee, H. & Schreur, A. (1985). Behaviour and impact assessment of heavy metals in estuarine and coastal zones, In: *Metals in Coastal Environments of Latin America*. U. Seeliger; L.D. de Lacerda & S.R. Patchineelam (Eds.), 157-198.

Schoonen, M.A.A. & Barnes, H.L. (1991). Reactions forming pyrite and marcasite from solution: II. Via FeS precursors below 100°C. *Geochimica et Cosmochimica Acta*, Vol. 55, pp. 1505-1514.

Shaw, T.J.; Gieskes, J.M. & Jahnke, R.A. (1990). Early diagenesis in differing depositional environments: the response of transition metals in pore water. *Geochimica et Cosmochimica Acta*, Vol. 54, pp. 1233-1246.

Stephens, S.R.; Alloway, B.J.; Parker, A.; Carter, J.E. & Hodson, M.E. (2001). Changes in the leachability of metals from dredged canal sediments during drying and oxidation. *Environmental Pollution*, Vol. 114, pp. 407-413.

Summers, J.K.; Wade, T.L. & Engle, V.D. (1996). Normalization of metal concentrations in estuarine sediments from the Gulf of Mexico. *Estuaries*, Vol. 19, No.3, pp. 581-594.

Taylor, S.R. (1964). Abundance of chemical elements in the continental crust: a new table. *Geochimica et Cosmochimica Acta*, Vol. 28, pp. 1273-1285.

Tessier, A. & Campbell, P.G.C. (1987). Partitioning of trace metals in sediments: relationships with bioavailability. *Hydrobiologia*, Vol. 149, pp. 43-52.

Tessier, A. & Campbell, P.G.C. (1988). Partitioning of trace metals in sediments, In: *Metal Speciation: Theory, Analysis and Application*, J.R. Kramer & H.E. Allen (Eds.), 183-199, Lewis Publishers, Inc.

Tessier, A., Campbell P.G.C. & Bisson, M. (1979). Sequential extraction procedure for the speciation of particulate trace metals. *Analytical Chemistry*, Vol. 51, pp. 844-851.

Turekian, K.K. & Wedepohl, K.H. (1961). Distribution of the elements in some major units of the Earth's Crust. *Geological Society of America Bulletin*, Vol. 72, pp. 175-192.

Uhrberg, R. (1982). Acid digestion bomb for biological samples. *Analytical Chemistry*, Vol. 54, pp. 1906-1908.

Ure, A.M.; Quevauviller, P.; Muntau, H. & Griepink, B. (1993). Speciation of heavy metals in soils and sediments: an account of the improvement and harmonization of extraction techniques undertaken under the auspices of the BCR of the Commission of the European Communities. *International Journal of Environmental Analytical Chemistry*, Vol. 51, pp. 135-151.

Varekamp, J.C. (1991). Trace element geochemistry and pollution history of mud flats and marsh sediments from the Connecticut River estuary. *Journal of Coastal Research*, Vol. SI11, pp. 105-124.

Vega, F.A.; Covelo, E.F & Andrade, M.L. (2008). Impact of industrial and urban waste on the heavy metal content of salt marsh soils in the southwest of the province of Pontevedra (Galicia, Spain). *Journal of Geochemical Exploration*, Vol. 96, pp. 148-160.

Vilas, F.; Bernabeu, A. M. & Méndez, G. (2005). Sediment distribution pattern in the Rías Baixas (NW Spain): main facies and hydrodynamic dependence. *Journal of Marine Systems*, Vol. 54, pp. 261-276.

Wangersky, P.J. (1986). Biological control of trace metal residence time and speciation: A review and synthesis. *Marine Chemistry*, Vol. 18, pp. 269-297.

Wedepohl, K.H. (1971). Environmental influences on the chemical composition of shales and clays, In: *Physics and chemistry of the Earth*, L.H. Ahrens; F. Press; S.K. Runcorn & H.C. Urey, (Eds.), 307-331, Oxford, Pergamon.

Wedepohl, K.H. (1991). The composition of the upper Earth's crust and the natural cycles of selected metals: metals in natural raw materials; natural resources, In: *Metals and Their Compounds in the Natural Environment*, E. Merian (Ed.), 3-17, Weinheim: VCH.

Wignall, P.B. (1994). *Black shales and their controversies*, Oxford University Press, 124 p.

Zoller, W.H.; Gladney, E.S.; Gordon, G.E. & Bors, J.J. (1974). Emissions of trace elements from coal fired power plants, In: *Trace Substances in Environmental Health*, D.D. Hemphill (Ed.), Vol. 8, 167-172, Rolla, University of Missouri, Columbia.

Zwolsman, J.J.G; Berger, G.W. & Van Eck, G.T.M. (1993). Sediment accumulation rates, historical input, postdepositional mobility and retention of major elements and trace metals in salt marsh sediments of the Scheldt Estuary, SW Netherlands. *Marine Chemistry*, Vol. 44, pp. 73-94.

Using a Multi-Scale Geostatistical Method for the Source Identification of Heavy Metals in Soils

Nikos Nanos[1] and José Antonio Rodríguez Martín[2]
*[1]School of Forest Engineering - Madrid Technical University
Ciudad Universitaria s/n, Madrid
[2]I.N.I.A. Department of the Environment, Madrid
Spain*

1. Introduction

For quite a long time, soil has been considered a means with a practically unlimited capacity to accumulate pollutants without immediately producing harmful effects for the environment or for human health. Presently, however, we know that this is not true. Public awareness has been raised on the harmful potential of some soil trace elements –commonly known as heavy metals- that can accumulate in crops and may end up in human diet through the food chain. Many studies have confirmed that heavy metals may accumulate and damage crops or even mankind (Otte et al., 1993; Dudka et al., 1994; Söderström, 1998). Along these lines, the most dangerous metals owing to their toxicity for human beings are Cd, Hg and Pb (Chojnacka et al., 2005).

Natural concentration of heavy metals in soil is generally very low and tends to remain within very narrow limits to ensure an optimum ecological equilibrium. Nonetheless, human activities that involve emitting large quantities of heavy metals into the environment have dramatically increased natural concentrations in the last century. Although soils are quite capable of cushioning anthropogenic inputs of toxic substances, there are times when this capacity is exceeded, which is when a pollution problem arises.

The natural concentration of heavy metals in soils depends primarily on geological parent material composition (Tiller, 1989; Ross, 1994; Alloway, 1995; De Temmerman et al., 2003; Rodríguez Martín et al., 2006). The chemical composition of parent material and weathering processes naturally conditions the concentration of different heavy metals in soils (Tiller, 1989; Ross, 1994). In principle, these heavy metals constitute the trace elements found in the minerals of igneous rocks at the time they crystallize. In sedimentary rocks, formed by the compaction and compression of rocky fragments, primary or secondary minerals like clays or chemical precipitates like $CaCO_3$, the quantity of these trace elements depends on the properties of the sedimented material, the matrix and the concentrations of metals in water when sediments were deposited. In general, concentrations of heavy metals are much higher in igneous rocks (Alloway, 1995; Ross, 1994). Nonetheless, these ranges vary widely, which implies that the natural concentration of heavy metals in soil will also vary widely.

Variability of heavy metals in soil is also associated with the variability of the physico-chemical properties of soil. pH, organic matter, clay minerals, metal oxides, oxidation-reduction reactions, ionic exchange processes, or adsorption, desorption and complexation phenomena, are the main edaphic chracteristics relating to retention of metals in soil. Nonetheless, some metals show a strong affinity to organic matter, while others have a strong affinity to clays, and to Fe and Mn oxides; yet they are elements which tend to precipitate in a carbonate form. Most of these soil properties will depend on the geological parent material. However, they will also be subject to not only the electric charge of metals in relation to the former saturation of other ions, but to all the interactions taking place naturally among the various edaphic parameters.

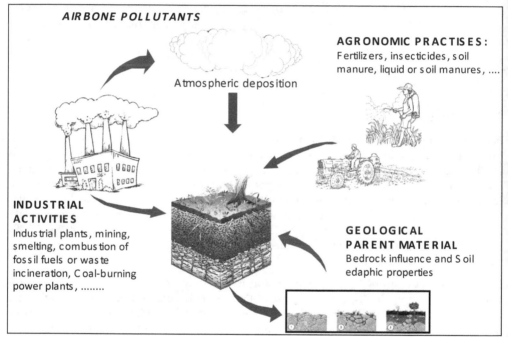

Fig. 1. Heavy metal input in soil

1.1 Anthropogenic influence on the content and distribution of heavy metals in soil

Many productive activities like mining, smelting, industrial activity, power production, pesticides production or waste treatment and spillage, represent sources of metals in the environment (Gzyl, 1999; Weber & Karczewska, 2004). The concentration of metals in soil can increase directly, or be due to the impact of atmospheric deposition (Figure 1) caused by proximity to industrial plants (Colgan et al., 2003) or through fossil fuel combustion (Sanchidrian & Mariño, 1980; Martin & Kaplan, 1998). Thus the atmosphere, which is an important means of transport for heavy metals originating from various emission sources, is the first factor that enriches soil in heavy metals. Although the natural entrance of metals to the atmosphere derives from volcanoes and the evaporation of the earth's crust and oceans, the main current input is of an anthropic origin. Airborne pollutants produced by mining, combustion of fossil fuels or waste incineration have significantly increased the emission of heavy metals into the atmosphere, which are subsequently deposited on the soil surface by deposition (Figure 2). The effects of this pollution can become evident in soil at a distance of hundreds of kilometers from the emission source (Hutton, 1982; Nriagu, 1990; Alloway & Jackson, 1991; Navarro et al., 1993; Engle et al., 2005).

In addition, normal agricultural practices may cause enrichment of heavy metals (Errecalde et al., 1991; Kashem & Singh, 2001; Mantovi et al., 2003). These practices are an important source of Zn, Cu, and Cd (Nicholson et al., 2003) due to the application of either liquid and soil manure (or their derivatives, compost or sludge) or inorganic fertilizers. Finally, there is also the pollution resulting from spillages, which are normally isolated and easy to identify, generally because they lead to very high values which multiply the expected soil content by several units. These sporadic sources of pollution are not usually observed in areas employed for growing crops.

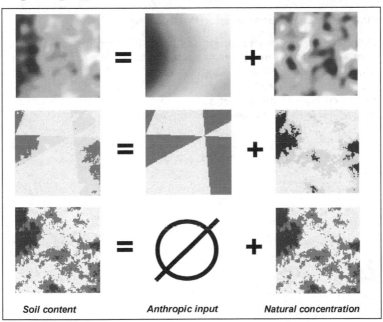

Soil content   *Anthropic input*   *Natural concentration*

Fig. 2. Schematic representation of multiple sources of heavy metals in soil. In the upper panels, we present the spatial distribution of a heavy metal enriched by airborne pollution – and subsequent deposition-. In the middle panels, we provide a hypothetical example of a heavy metal enriched by agricultural practices such as fertilization (green indicates input of heavy metals). In the lower panels, natural soil content is assumed to not be enriched by any human activity. Note that in real case studies we may only observe the sum of two inputs.

1.2 Identification of sources of soil heavy metals

Source identification and apportionment of metal elements in soil are not straightforward. Quantities of metals introduced into soil through industrial activities, or any other human activity, do not provide any trace of their anthropogenic origin and, once inside soil, they behave like any other similar natural analogue which is already in soil. Likewise metal like copper, which forms part of soil after fertilizing farmland, shows no distinguishing element of its origin. This makes us wrongly think that human input does not exist if the total concentration in soil does not reach levels considered to be polluting. Consequently, what we observe in the analysis of these elements in soil tends to result from summing the two inputs (Figure 2).

Despite the difficulty of separating human input from natural input, the separation task is greatly needed for correct edaphic resource management and to prevent its pollution. In this chapter, we present a multiscale geostatistical method known as a factorial kriging analysis which -under certain circumstances- can provide a mathematical framework to distinguish natural soil enrichment from that of an anthropogenic kind in heavy metals. The method was initially presented by Matheron (1982) and has been used repeatedly in soil science (Goovaerts, 1992, Castrignano, 2000, Rodriguez et al.., 2008).

1.3 Description of the study areas used in the analysis

The information used in this chapter originates from the sampling and the analysis of the two most important hydrographic basins in Spain (Figure 3) in a study conducted to obtain reference values in Spanish soils (Rodríguez et al., 2009). They all present different lithologies and geologies, as well as common edaphic processes. Likewise, they also reflect possible inputs from both farming treatments and industrial activities which modify the contents and distribution of these elements in each valley.

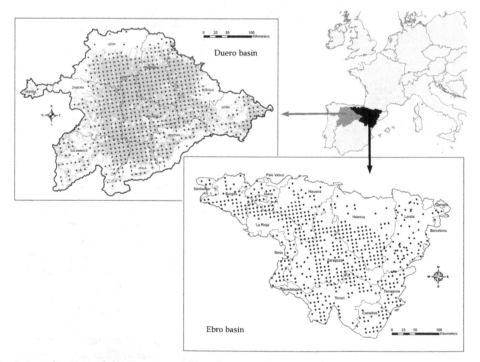

Fig. 3. Localization of the Duero and Ebro basins. Soil samples are plotted over the two river basins

1.3.1 The Duero basin

The Duero river basin is the largest of its kind in the Iberian Peninsula, with a total surface of 97,290 km2, of which 78,954 km2 are found in Spanish territory (Confederación Hidrográfica del Duero, www.chduero.es). From a geological point of view, the hydrographic Duero basin consists of a well-defined geological unit, the Duero depression and its borders. The Duero Basin is an intraplate continental basin which developed from the Late Cretaceous to the Late Cenozoic. Over this time span, the basin acted as a foreland basin of the surrounding Cantabrian Zone and the Basque-Cantabrian Range in the north, the Iberian Range in the east and the Central System in the south, which constitute complex fold and thrust belts. These Alpine compressional ranges, constituted by Palaeozoic and Mesozoic rocks, are thrusting, in many cases, the Cenozoic deposits of the Duero Basin, which are virtually undeformed (Gomez et al., 2006). The basin is filled mainly with siliciclastic sediments on the margins and evaporites in central areas, showing an endorrheic arrangement (Tejero et al., 2006). The basin's general facies distribution corresponds to a continental foreland basin model with alluvial fan deposits grading into alluvial plains, and evaporitic and carbonated lacustrine environments toward the centre of the basin (Gomez et al., 2006).

The total population in the basin is around 2,210,541 inhabitants (Municipal Register, 2006), which has barely varied in the last hundred years. The Spanish Autonomous Community of Castilla y León is one of the main cereal-growing areas in Spain. Apart from pulses, such as carob and chickpea, the practice of growing sunflowers has extended in the southern countryside. However, the number of cultivated vineyard hectares (56,337 ha) lowered considerably in the last three decades of the past century.

1.3.2 The Ebro basin

The Ebro Valley, in the northeast region of the Iberian Peninsula, is framed by three mountain ranges: the Pyrenees to the north, the Iberian Chain to the southwest, and the Catalonian Coastal Ranges to the southeast. The structural development of these ranges controlled the evolution of this basin in tectonic and structural terms, and as regards stratigraphic and sedimentologic aspects. The Pyrenean range is a fold and thrust belt that developed during the Tertiary (the following 60 million years) as the Iberian block converged toward the European Plate. Its metamorphic core marks the border between France and Spain, with foreland structures verging into both countries. The tertiary stratigraphic units also increase in thickness northwardly. The tectonic load of the alochtonous units and the formation of a cold lithospheric root during Pyreneean shortening induced the deflection of the Ebro basin (Brunet, 1986, Roure et al., 1989). The depth of the Tertiary basin increases northwardly, reaching values of 4000 m beneath the sea level below the Pyrenees (Riba et al., 1983). The most exposed rocks within the basin area are of the Oligocene-Miocene age (including clastic, evaporite and carbonate facies) and of an alluvial and lacustrine origin (Riba et al., 1983; Simon-Gomez, 1989). During the Quaternary, incision of the drainage system caused the isolation of structural platforms, the tops of which are composed of near-horizontal Neogene limestones. Contemporaneously, several nested levels of alluvial terraces and sediments developed (Simón and Soriano, 1986). The Quaternary levels comprise mainly gravels, sand and slits.

The Ebro river region, with a population of around 3.25 million, is intensively industrialized. The Ebro river region is also an important agricultural area in Spain, with 4.2 million ha (Fig. 1) of agricultural topsoil (the total basin area is 9.5 million ha).

1.3.3 Soil samples and chemical analysis

The sampling scheme was based on an 8x8 km grid. Samples were located using GPS and topographics maps on a scale of 1:25,000. Each sample was defined as a composite made up of 21 subsamples collected with the Eijkelkamp soil sampling kit from the upper 25 cm of soil in a cross pattern. Further details can be found in Rodríguez et al. (2009).

Soil samples were air-dried and sieved with a 2 mm grid sieve. Soil texture was determined for each sample. After shaking with a dispersing agent, sand (2 mm–63 mm) was separated from clay and silt with a 63 μm sieve (wet sieving). A standard soil analysis was carried out to determine the soil reaction (pH) in a 1:2.5 soil-water suspension (measured by a glass electrode CRISON model Microph 2002) and organic matter (%) by dry combustion (LECO mod. HCN-600) after ignition at 1050°C and discounting the carbon contained in carbonates. Carbonate concentration was analyzed by a manometric measurement of the CO_2 released following acid (HCl) dissolution (Houba et al., 1995).

Metal contents (Cr, Ni, Pb, Cu, Zn, Hg and Cd) were extracted by aqua regia digestion of the soil fraction in a microwave (ISO 11466, 1995). Heavy metals in soil extracts were determined by optical emission spectrometry (IPC) with a plasma spectrometer ICAP-AES. Mercury in soil extracts was determined by cold vapor atomic absorption spectrometry (CVAAS) in a flow-injection system. The summary statistics of soil parameters and heavy metal contents are listed in Table 1.

	Duero basin 721 soil samples					Ebro basin 624 soil samples				
	Mean	Median	S.D.	1st Qu	3rd Qu	Mean	Median	S.D.	1st Qu	3rd Qu
S.O.M	1.74	1.3	1.494	0.88	2.08	2.24	2	1.41	1.4	2.6
Soil pH	7.19	7.7	1.234	6	8.3	8.04	8	0.59	8	8.4
E.C	0.18	0.15	0.16	0.1	0.21	0.59	0.27	0.85	0.21	0.45
CaCO3	9.41	3	14.45	1	10.1	29.68	31	16.08	20	40
Sand	59.2	61	19	46	75	38.67	38	17.09	26	50
Silt	20.9	19	12.7	10	31	22.05	21	8.7	16	27
Clay	19.9	19	10.6	12	26	39.41	39.4	13.08	30	48
Cr	20.53	18	14.9	10	27	19.82	18	11.18	13	24
Ni	15.08	13	9.99	8	20	19.26	18	8.64	13	23
Pb	14.06	13	6.79	9	17	16.98	15	7.63	12	20
Cu	11.01	10	7.84	6	15	16.68	13	11.89	10	21
Zn	42.42	38	23.01	27	53	57.4	55	24.27	41	69
Hg	42.05	30	58.43	18	50	33.83	27	38.66	16	44
Cd	0.159	0.1	0.14	0.07	0.2	0.413	0.4	0.159	0.3	0.5

1st Qu, 3rd Qu, first and third quartile; SD, standard deviation ; SOM, soil organic matter (%); CaCO3, carbonates (%) ; EC, Soil electrical conductivity (dS m-1)

Table 1. Statistical summary of metal concentrations of soil (in mg/kg for Zn, Fe, Cu, Cd and μ/kg for Hg) and some soil properties.

2. Factorial kriging analysis

2.1 Univariate analysis

Usually physical and chemical soil variables are intrinsically structured around more than one scale of variation, and it is up to the researcher to identify the most important one(s) for his/her study. In factorial kriging -and in geostatistics in general- identification of the scale (or scales) of variation of a variable is done via the variable's sample (or experimental) variogram.

Let us consider a regionalized variable sampled at N points $\{ x_a : a = 1,...,N \}$ such as the Cd concentration in Ebro basin top soil. The sample variogram for Cd is then computed according to the formula:

$$\hat{\gamma}(h) = \frac{1}{2N(h)} \sum_{x_\beta - x_\alpha \approx h} \left[z(x_\beta) - z(x_\alpha) \right]^2 \tag{1}$$

where $N(h)$ is the number of pairs of data locations separated by, approximately, distance h and $\hat{\gamma}(h)$ the semivariance for lag h. In fact, the sample variogram is constructed by grouping pairs of observations into several discrete distance classes or lags. The average separation distance of all the pairwise points falling within lag h is plotted against half the average semivariance for each lag computed with formula 1, giving raise to a scatter plot similar to that depicted in Figure 4. Typically, semivariance exhibits an ascending behavior near the origin (h=0), whereas at longer separation distances, it levels off at a maximum value called the variogram sill. The distance at which the sill is reached is called the variogram range, while the term nugget is used for the semivariance value at a distance of h=0.

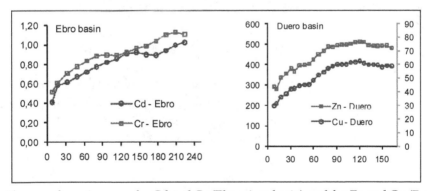

Fig. 4. The sample variograms for Cd and Cr (Ebro river basin) and for Zn and Cu (Duero basin). Horizontal distance is measured in km; vertical axes project the semivariance for a certain distance lag. Note also that the semivariance values for the Ebro have been standardized to unit variance (this is not the case for Duero basin). Note too how the slope of the sample variograms changes before reaching the sill, indicating possible variation on several spatial scales.

Sample variograms, like those shown in Figure 4, indicate that the variable is not distributed randomly in space, rather it is spatially correlated -with an extended spatial correlation equal to the range of the variogram- so that the pairs of observations separated by a distance shorter than the range of the variogram are more similar on average than usual. When sample variograms also show two or more distinct slopes while ascending toward the sill, the variable can be considered a multi-scale distributed variable. In the Ebro case study, for instance, the sample variograms for Cd and Cr exhibit two different slopes before the sill. More specifically, Cd presents a steep slope in the first distance lag (20 km), a second one up to approximately 100 km, and a last slope change until ca. 220 km. Cr, on the other hand, exhibits two slopes at approximately 90 km and 200 km. Conversely, the sample variograms for Zn and Cu in the Duero basin seem to reach the variogram sill at distance of 100 km. Zinc presents a considerable step ascension in its semivariance at short distances of less than 20 km.

2.1.1 Variogram modeling in the univariate case

After the sample variogram has been calculated, it is modeled using either a unique variogram model or a combination of more than one variogram models (also called structures). The variables presenting spatial variation on a unique spatial scale are conveniently modeled using a combination of two variogram models: a nugget model and a structure with a range of spatial correlation equal to the range of the sample variogram. However, whenever multiscale variation is present in the sample variogram, models with more structures should be adopted for modeling. Several variogram models can be used at this stage (a detailed list of variogram models and their characteristics can be found in Chilés and Delfiner (1999)). However, for multiscale variation, modeling practice has shown that spherical models are the most convenient. If we consider that $k=1,...,q$ denotes the number of structures used to model the experimental variogram, so the variogram model (also called the linear model of regionalization) can be written as:

$$\gamma(\mathbf{h}) = \sum_{k=1}^{q} \gamma_k(\mathbf{h}) = \sum_{k=1}^{q} b_k g_k(\mathbf{h}) \qquad (2)$$

where b_k is the partial sill and $g_k(\mathbf{h})$ is the variogram model of the kth structure.

A very important property of the linear model of regionalization is that it can be used to decompose the original random function $Z(x)$ into q independent random functions (called spatial components) corresponding to spatial scale k. Decomposition is based on the following model:

$$Z(x) = \sum_{k=1}^{q} a_k Y_k(x) + m \qquad (3)$$

where a_k are known coefficients and $Y_k(x)$ are the orthogonal spatial components with spatial covariance $c_k(\mathbf{h})$:

$$E\{Z(x)\} = m$$

$$E\{Y_k(x)\} = 0$$

$$Cov\left\{Y_k(x), Y_{k'}(x+\mathbf{h})\right\} = \begin{cases} c_k(\mathbf{h}) & \text{if } k=k' \\ 0 & \text{otherwise} \end{cases}$$

This decomposition is of important practical interest: it makes the distinction of the original random function $Z(x)$ into q uncorrelated random functions possible, which represent different scales of variation. Additionally, the estimated spatial components can be mapped using a modified kriging system of equations, which may assist in their interpretation (see also the example in the next Chapter). Mapping spatial components is done with a modified kriging system of equations. Given a set of n observations, the optimal weights $\left(\lambda_\beta\right)$ for the estimation of spatial component k are given by the solution of the following system of equations (ISATIS, 2008):

$$\begin{cases} \sum_{\beta=1}^{n} \lambda_\beta \gamma_{\alpha\beta} + \mu = \gamma_{a0}^{k} \\ \sum_{\beta=1}^{n} \lambda_\beta = 0 \end{cases} \qquad a=1,....,n \qquad (4)$$

where $\gamma_{\alpha\beta}$ is the semivariance between data locations a and β, and γ_{a0}^{k} is the semivariance between location a and the point where an estimation is required. System [4] is identical to a usual kriging system of equations, save the term γ_{a0}^{k}, which is the semivariance computed using only the variogram model corresponding to the kth structure. Another difference is that kriging weights λ_β must sum to zero (unlike ordinary kriging where weights sum to unity). This difference is due to the fact that the mean of the random function $Z(x)$ is considered a part of the spatial component with the largest range [max(k)]. In this case (i.e., when estimating the largest range component) the system [4] should be rewritten in order to account for the varying spatial mean of the function (Isatis, 2008):

$$\begin{cases} \sum_{\beta=1}^{n} \lambda_\beta \gamma_{\alpha\beta} + \mu = \gamma_{a0}^{\max(k)} \\ \sum_{\beta=1}^{n} \lambda_\beta = 1 \end{cases} \qquad a=1,....,n \qquad (5)$$

2.1.2 Example: A model of regionalization for zinc in the Duero basin

The sample variogram for Zn concentration in Duero basin soil samples presents two different slopes before reaching the sill at a distance of approximately 120 km (Figure 5). Apparently, the total spatial variability of this variable is structured around two spatial scales at 20 km (local) and at 120 km (regional). The model of regionalization adjusted to the sample variograms of Zn in the Duero Basin is composed of three structures, namely a nugget effect model and two spherical models, with 20 km and 120 km ranges of spatial correlation. The spatial components corresponding to this model may now be used to decompose the original variable into two components by separating the variation observed on a small spatial scale from the one on a larger scale (see the maps in Figure 5).

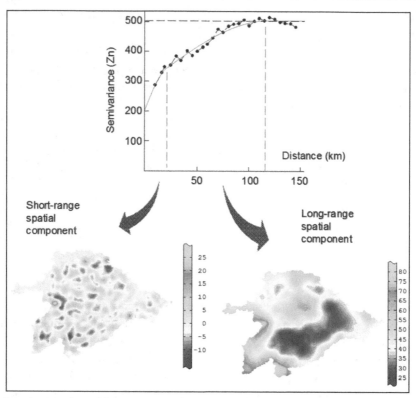

Fig. 5. Mapping two spatial components for Zn concentration in Duero basin soils. The upper panel shows the linear model of regionalization adjusted to the experimental variogram. The lower panels present the estimated spatial components with short- (left panel) and long- (right panel) range variation.

The variation of the component with the longest range (120 km in this case) in the distribution of heavy metals in soils tends to attribute to geological-type factors of a natural origin. Rock decomposition and its input of elements to soil formation (Alloway, 1995) are two main factors that influence on this scale. Hence, large lithological units determine zinc distribution on this scale in the Duero valley. On the other hand, zinc is not a metal related with atmospheric deposition processes over long distances, and is not associated with an industrial activity that might influence the Duero valley on such a large scale. An alternative interpretation of the influence of farming treatments on zinc content in soil is limited to the extension of plots used for crop-growing. The spatial extent of the spatial component in Figure 5 is much greater than the extended land use polygons in the study area.

Unlike the long-range component, its counterpart has a range of spatial correlation of just 20 km, which may be attributable to both antropogenic and natural factors. To a minor spatial extent, zinc in soil would be related with edaphic parameters, such as organic matter, clays, cationic exchange capacity, or presence of oxides of other metals like Fe or Al, which favor its metals retention and whose variation in soil may correspond to the ranges of this scale. On the other hand, the variability on this small-range scale (20 km) may also reflect antropogenic activities in either agriculture or small industrial plants. With this kind of statistical analysis, it is impossible to distinguish the repercussion or influence of both factors (human input from natural input) beyond assumptions. However, and in relation to some low Zn contents (save the odd exception), it would be possible to rule out that inputs originating from generic industrial activities, spillages or sporadic pollution sources would be of much influence on this scale. Nonetheless, it would be necessary to know the relationship with other edaphic parameters, specifically with those metals that might relate more with farming activities.

2.2 Multivariate analysis

The natural extension of the variogram to include two random functions (i and j) is called the cross-variogram between Z_i and Z_j, which is estimated as:

$$\hat{\gamma}_{ij}(h) = \frac{1}{2N(h)} \sum_{x_\beta - x_\alpha \approx h} \left[Z_i(x_\beta) - Z_i(x_\alpha) \right]\left[Z_j(x_\beta) - Z_j(x_\alpha) \right] \tag{6}$$

where all the symbols are as in (1) but using the subscript i or j to distinguish between the two variables.

Variogram modeling is more complicated in the multivariate case since one needs to model a total of $p(p+1)/2$ direct and cross-variograms (where p is the number of variables). Multivariate variogram fitting is usually accomplished with the use of multiple nested structures where each one corresponds to a different scale of variation. The model –called the linear model of coregionalization, LMC- can be written as:

$$\gamma_{ij}(\mathbf{h}) = b_{ij}^{(1)} g^{(1)} + b_{ij}^{(k)} g^{(k)}(\mathbf{h}) + ... + b_{ij}^{(q)} g^{(q)}(\mathbf{h}) \quad \forall \; i, j \tag{7}$$

where $\gamma_{ij}(\mathbf{h})$ is the variogram model for variables i and j (for $i=j$ the auto-variogram is obtained), $b_{ij}^{(k)}$ is the partial sill for the ijth variogram for structure k, while $g^{(k)}(\mathbf{h})$ represents the type of variogram model (i.e., exponential, spherical, etc.) for structure k. The first structure $g^{(1)}$ represents the nugget effect model.

The flexibility of a specific LMC (that is, its ability to model a set of experimental variograms) is based on the total number of basic structures (q) and their corresponding range of spatial correlation, as well as the partial sill $b_{ij}^{(k)}$ for each variogram model and structure. Partial sills $b_{ij}^{(k)}$ may vary across variograms (under some restrictions; see Wackernagel (1998)), but the range of spatial correlation of each structure $g^{(k)}(\mathbf{h})$ should be the same for the set of $p(p+1)/2$ variogram models. If we arrange the partial sills $b_{ij}^{(k)}$ of the LMC in a matrix form, we obtain the so-called co-regionalization matrices. The coregionalization matrix for structure k is a positive semi-definite symmetric ($p \times p$) matrix with diagonal and off-diagonal elements of the partial sill of the auto- and cross-variogram models, respectively, obtained from the LMC:

$$\mathbf{B}_k = \begin{bmatrix} b^{(k)}_{11} & \cdots & b^{(k)}_{1p} \\ \cdots & \cdots & \cdots \\ b^{(k)}_{p1} & \cdots & b^{(k)}_{pp} \end{bmatrix} \tag{8}$$

Note that the LMC is a permissible model, but only when all the co-regionalization matrices $\mathbf{B}_k$ are positive definite. This constraint makes multivariate variogram fitting a difficult task. Should only two variables be implicated (p=2), the linear model of coregionalization can be fitted manually. However for p>2, a weighted least squares approximation is needed which offers the best LMC fit under the constraint of positive semi-definiteness of $\mathbf{B}_k$ (Goulard and Voltz 1992). However, the use of the LMC facilitates multivariate variogram modeling since it reduces the problem of fitting a total of $p(p-1)/2$ variogram models to their experimental counterparts to the problem of deciding the total number (q) of structures to use, as well as the type (i.e., spherical, exponential, etc.) of each structure. This is a central decision in the MFK framework. There are no general rules to guide this decision; however, some recommendations may prove useful (see also Goulard and Voltz (1992)):

- Experimental variograms are the basis for deciding the number and range of elementary structures to be used for modeling. One usually tries to find subsets of variables that have similar variogram characteristics (equal range of spatial correlation).
- In practice, it is difficult to distinguish more than three structures (a nugget plus two other structures) in a set of experimental variograms. Practical experience has shown that three basic structures are sufficient for modeling a large number of variables.
- Usually some of the original variables in the dataset should share the same structures.
- Specifically for heavy metal distribution in soils, geological maps can prove most helpful for deciding the number and ranges of the spatial correlation of individual model structures. The spatial distribution of heavy metals in soil has been observed to be in close relationship with the basic geological features –and their spatial extent- of the study area.

2.2.1 Example: Fitting the linear model of coregionalization in the Ebro river basin

Let us now describe the procedure for fitting a linear model of corregionalization for the spatial distribution of heavy metals in the Ebro basin (Rodriguez et al., 2008). By looking at the direct experimental variograms (Figure 6), we note a change in their slope at a distance of approximately 20 km, which is more prominent for some metals (Hg and Cd) than for others (Zn). This common feature in all the sample variogram indicates that the linear model of coregionalization should include a short-range structure. Note also that all the sample variograms seem to reach the sill value at a distance of approximately 200 km, a fact which also makes necessary the inclusion of a long-range component in the model of coregionalization to be built. Finally, we note that the ascension in the semivariance observed between the shortest range (20 km) and the longest one (200 km) is not linear; instead we observe a stabilization at distances of approximately 100 km.

The features observed in the sample variograms act as a basis for deciding that the linear model of coregionalization to be postulated -and fitted afterward to the sample variograms-should be composed of at least three structures with ranges of approximately 20 km, 100 km and 200 km (plus a nugget effect model which is present in all the sample variograms). Cross variograms - not presented here - were also taken into account when building the model. However, the postulated model should be chiefly designed to fit the direct rather than the cross-variograms as precisely as possible.

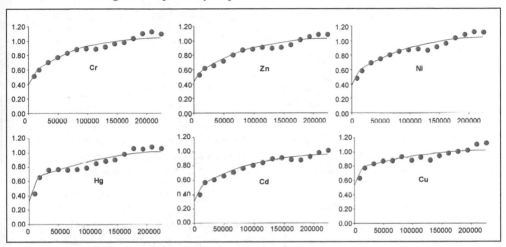

Fig. 6. Six of the seven variogram models of the linear model of coregionalization for the concentration of heavy metals in Ebro basin soils. Horizontal axes units are provided in meters, while vertical axes represent the standardized –to unit sill- semivariance of the corresponding element.

In Figure 6, we graphically present the linear model of coregionalization fit after an iterative procedure included in the Isatis software (Isatis, 2008). The fit is not actually perfect, especially for some elements such as Hg. Yet when considering a multiple variogram model, a compromise between model simplicity and accuracy of fit should be reached.

The number and range of the structures adopted in this model are coherent from a practical viewpoint. Note that within the study area, we have reported the presence of several industrial activities that may potentially enrich the basin soil through airborne pollution and subsequent deposition. We expect that this activity may have altered the spatial distribution at intermediate distances from the point source. Mercury emitted from industrial plants, for instance, has been reported to travel tens of thousands of kilometers before being deposited in soil. On the other hand, large geological features within the study area make us think that natural geological variation could be responsible for the spatial distribution of some heavy metals on a large spatial scale (that with the longest range). Finally, differences in land use and short-range scale natural processes affecting pedogenesis are believed to act on short spatial scales, and are presumably responsible for the short-range scale effects in the distribution of soil heavy metals.

In a very similar way, we built the linear model of coregionalization for heavy metals concentration in Duero basin soils. The sample variograms features in this basin are similar: variograms present a short-range structure at a distance of 20 km and a medium-range structure at 100 km – 120 km. A substantial difference, however, between the two basins is noted in the absence of long-range variation in the Duero basin. Therefore, the long-range structure (220 km) found in the Ebro region is absent in this model (see also Figure 7).

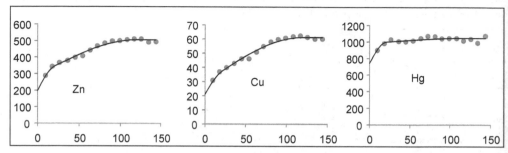

Fig. 7. Three of the seven variogram models of the linear model of coregionalization for the heavy metals concentration in Duero basin soils. Horizontal axes units are provided in kilometers, while vertical axes represent the semivariance of the corresponding element.

2.2.2 Extracting spatial components

Analogically to the univariate case, the LMC permits the decomposition of the original random functions $Z_i(x)$ into a linear combination of the q mutually uncorrelated random functions Y_{lk}, called regionalized factors (RFs):

$$Z_i(x) = m_i(x) + \sum_{k=1}^{q}\sum_{i=1}^{p} a_{il}^k Y_{lk}(x) \qquad i=1,...,p$$

or in matrix form:

$$\mathbf{Z} = \mathbf{m} + \sum_{k=1}^{q} \mathbf{A}_i \mathbf{Y}_k \qquad\qquad (9)$$

where $m_i(x)$ is the varying mean of the function and $\mathbf{A}_k$ is a matrix of unknown coefficients. In fact, only the coregionalization matrix for the kth structure:

$$\mathbf{B}_k = \mathbf{A}_k \mathbf{A}'_k \qquad\qquad (10)$$

can be estimated. Matrix $\mathbf{A}_k$ is not uniquely defined since there is an infinitive number of $\mathbf{A}_k$ which satisfies Equation (10). However, a principal components analysis (PCA) can provide a natural determination of matrix $\mathbf{A}_k$ (Chilés and Delfiner 1999). More specifically, a PCA is applied to each coregionalization matrix $\mathbf{B}_k$ separately, to provide a set of p eigenvalues and their corresponding eigenvectors $(\mathbf{u}_l)$:

$$\mathbf{B}_k = \mathbf{Q}_k \mathbf{\Lambda}_k \mathbf{Q}'_k \qquad\qquad (11)$$

where $\mathbf{Q}_k$ is the orthogonal matrix of eigenvectors and $\mathbf{\Lambda}_k$ is the diagonal matrix of eigenvalues for spatial scale k.

Matrix $\mathbf{A}_k$ can then be estimated since:

$$\mathbf{A}_k = \mathbf{Q}_k \mathbf{\Lambda}_k^{1/2} \tag{12}$$

Using decomposition (9) and the estimation of matrix $\mathbf{A}_k$ from (12), one can decompose the p original variables into a set of uncorrelated RFs. Note that these factors have some remarkable properties:

- They are linear combinations of the p original variables.
- The first factors, corresponding to the largest eigenvalues, account for most of the variability observed in the dataset. Therefore, it is possible to reduce the dimensionality of the data and to visualize the phenomenon.
- It is possible that the resulting RFs represent some common mechanisms underlying the spatial distribution of the original variables. Hopefully, RFs can have meaningful interpretations (analogically to a classical PCA) .
- RFs remain uncorrelated at any separation distance $\mathbf{h}$ and not just at the same location (Chilés and Delfiner 1999). A classical PCA provides uncorrelated factors, but only for the separation distance $\mathbf{h}=0$.
- Perhaps the most remarkable property is that RFs can be constructed for each spatial scale separately. Therefore, by incorporating the spatial correlations revealed by the variograms, they can reveal mechanisms that act on different scales and that control the spatial distribution of the studied attributes.

2.2.3 Interpreting the regionalized factors: The circle of correlation

Interpretation of an RF is not always feasible given that RFs are determined using statistical and not ecological/physical criteria (maximize variance under the constraint of orthogonality). Nevertheless, the interpretation of RFs is of crucial importance for the analysis since it will be the basic tool for determining the exact physical mechanisms acting on different spatial scales and for controlling the spatial distribution of the studied variables.

The interpretation of the meaning of an RF can be assisted by all the well-known tools of a classical PCA, such as "scree" plots, loadings, and especially by the correlation between the regionalized factors and the regionalized variables, which is computed by:

$$\rho_{il} = q_{il} \sqrt{\frac{\lambda_l}{\sigma_i^2}} \tag{13}$$

where q_{il} is the loading of the ith variable for the lth principal component, λ_l is the variance of the lth RF and σ_i^2 the variance of the ith regionalized variable.

The pair of correlations of a regionalized variable with two RFs (usually the first two, which account for most of the variability) is plotted on a graph, called the circle of correlation (Saporta, 1990). When a variable is located near one of the two axes of the graph and away from the origin, it is well correlated with that specific RF and much of its variance is explained by the RF.

2.2.4 Example: Multiscale correlations among heavy metals in the Ebro river basin

In this Chapter, we use the linear model of corregionalization to decompose the initial multivariate set of heavy metal concentrations. Then we derive a new set of composite regionalized factors for several scales of variation. Finally, we employ a factorial kriging analysis results in an attempt to decompose the original dataset into a few regionalized factors.

For the Ebro river, three spherical models with ranges of 20 km, 100 km and 220 km were used. The three circles of correlation shown in Figure 8 represent heavy metal associations for the three scales of variation used in the linear model of corregionalization for soil samples. The association of different metals substantially differs if we compare the three circles of correlation. This indicates that, indeed, the correlation among soil elements depends on the spatial scale considered and that, conclusively, multiscale correlation is present in the study area. Grouping metals in the circle of each scale tends to reflect the influence phenomena that are common for all grouped elements. Changes made to the grouping of these metals when amending the observation scale also implies a change in the dominant factor which influences these elements. Note that in the absence of multiscale correlation, we expect the elemental associations to remain unchanged when moving from one spatial scale to another.

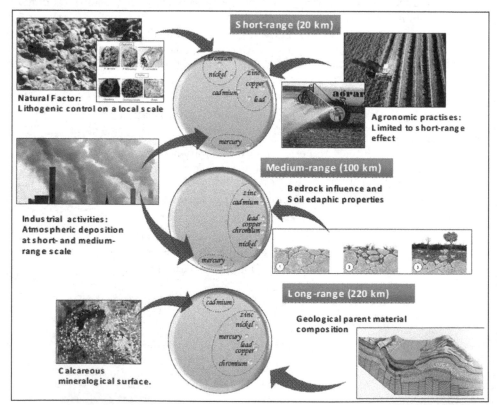

Fig. 8. Correlation circles for the concentration of seven heavy metals in the Ebro river basin. Correlation circles correspond to the three structures used to build the linear model of corregionalization.

The association between Hg and other elements is the most representative example of a correlation dependent on scale. Note that Hg seems to be an isolated element in the correlation circles for the short- and medium-range correlation circles, but seems to be associated with other elements when looking at the long-range correlation circles. Obviously, there should be a factor that acts on the smallest scale of variation, but not on the largest one, which makes mercury distribution different from that of the other elements. In general, mercury accumulations in soils are associated with atmospheric deposition (Engle et al., 2005). Anthropogenic emission of Hg represents about 60–80% of global Hg emissions. Mercury is an extremely volatile metal which can be transported over long distances (Navarro et al., 1993). In the Ebro valley, soil enrichment by atmospheric mercury covers small (20km) to medium (100km) scales. Not surprisingly, soil enrichment in mercury is not observed on the largest (among those studied) spatial scales, so man-made alteration of soil Hg has not become evident –at least not yet- on this scale.

Cd is another interesting element which shows correlations with other elements that change across scales. In the largest scale of variation (220 km), we may denote Cd as an isolated element in the circle of correlation which is, however, not observable on smaller spatial scales. Unlike Hg, however, this change can be attributed to natural factors. More specifically, Cd is the only element that tends to accumulate in calcareous soils (Boluda et at., 1988). Cadmium is adsorbed specifically by crystalline and amorphous oxides of Al, Fe and Mn. Metallic (copper, lead and zinc); alkaline earth cations (calcium and magnesium) particularly reduce Cd adsorption by competing for available specific adsorption and cation-exchange sites (Martin and Kaplan, 1998). Therefore, the isolated occurrence of Cd in the long-range correlation circle is due to the spatial distribution of the calcareous mineralogical surface generating major Cd accumulation.

On the local spatial scale (20km), association of Zn, Cu, Pb and Cd results concentrations after fertilization of arable soils, or pesticides and fungicides use related to crop protection. The association of four elements is caused by common agricultural practices, which may also prove evident on a small scale. Copper and zinc (Mantovi et al., 2003) increase through such practices especially copper which has been used as a pesticide form (copper sulfate) in viticulture. Zinc and cadmium concentrations increase through use of fertilizers. It is estimated that phosphated fertilizers make up more than 50% of total cadmium input in soils (de Meeûs et al., 2002). On the other hand, natural phenomena act in a short range, and variability of Cr and Ni is associated with the mineralogical structure of the study area (basic and ultramafic rock). Generally, anthropic inputs of Cr and Ni in fertilizers, limestone and manure are lower than the concentrations already present in soil (Facchinelli et al., 2001).

2.2.5 Example 2: Multiscale correlations among soil heavy metals in the Duero river basin

The linear model of co-regionalization in the Duero basin is developed from two structures, with scale ranges between 20 km and 120 km. Unlike the Ebro valley, no higher scale of variation is noted, which is around 200 km, despite having a similar surface. The Duero valley can be considered more homogeneous in terms of all the aspects that can influence the distribution of heavy metals in soils, and reducing spatial variation structures may well reflect this. Moreover, the Duero depression is shaped as a basin with tertiary and quaternary sediments of a continental origin. However, Paleogene materials of variable extensions outcrop from among the tertiary sediments, although they mainly limit the depression on the basin's edges. Variability from a geological viewpoint is inferior to variability in the Ebro. Yet from the farming perspective, crops do not present much diversity as dominant crops are cereals and vineyards. As mentioned in Section 1.4.4, the Duero valley is one of the main cereal-growing areas in Spain.

On the short-range scale (20km), only two groups of metals are seen, which are expressed as mercury isolation and a large series with the remaining elements (Figure 9). The main difference found with the Ebro valley lies in chrome and nickel grouping with metals like copper or zinc, which relate to farming practices. Although natural and inorganic fertilizers contain small amounts of both chrome and nickel, they do not significantly increase the soil natural content. The geological influence of nickel and chrome content has been clearly demonstrated (Alloway, 1995). The highest nickel concentrations are found in ultrabasic igneous rocks (peridotites, dunites and pyroxenites), followed by basic rocks (gabbro and basalt). Acid igneous rocks present lower chrome or nickel contents. Sedimentary rocks are especially poor in Cr or Ni. On the other hand, the edaphic parameters that influence heavy metals content in soil (texture, organic matter, etc.) evidence their influence on this scale. Soils with a thick and sandy texture contain less Cr or Ni than clayey soils, which is also a common process for the remaining metals (Cu, Pb, Cd and Zn). As expected, low heavy metal contents are also associated with low organic matter contents.

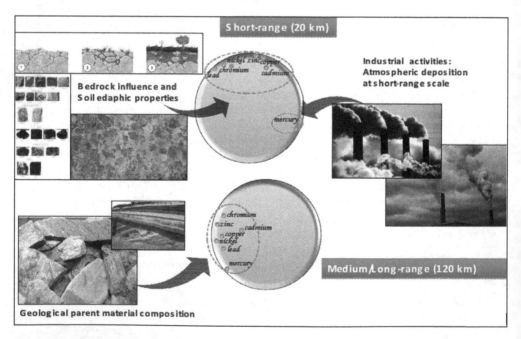

Fig. 9. Correlation circles for the concentration of seven heavy metals in the Duero basin. Correlation circles correspond to the two structures used to build the linear model of corregionalization.

On the medium-/long-range scales (120km), all heavy metals are grouped in the circle of correlation (Figure 9), which is the result of a common source of variability. On this scale, only large lithologies can influence the distribution of heavy metals in soils which, for the River Duero basin, derive mainly from the sedimentary materials deposited in its interior. However, the coordinated distribution of all metals, including Hg, on this scale evidences an edaphogenic process.

In conclusion, human activity is only shown on a small scale (20km) and is motivated by mercury inputs. As mentioned earlier, atmospheric deposition is the main way that mercury enters soil. Besides, on this small scale, it is probably a case of mercury being associated with ash and soot particles, which are deposited near pollution sources, such as those originating from coal power stations, or from heating systems to a lesser extent. Basically, no influence of farming treatments is observed. This aspect is linked to some relatively low levels of metals in soil, such as Cu, Cd or Zn. The dominant factor in the content and distribution of heavy metals in the Duero is natural, and is influenced by soil's physico-chemical properties in a short range and, to a greater extent, by geological parent material composition.

2.2.6 Spatial estimation of regionalized factors

The final step in factorial kriging analysis consists in mapping composite regionalized factors. Estimation is done with a modified cokriging step technique. Note that ordinary cokriging provides a spatial estimation of the primary variable using data for the primary variable and for one or more secondary variables. Typically, the primary variable is sampled over a limited number of points, while secondary data are more densely sampled (Wackernagel 1998). Estimation of regionalized factors is done by means of a modified cokriging system of equations, where measurements of the primary data are not available, and the cross covariance between the regionalized factor and the regionalized variables cannot be inferred directly. However, the spatial estimation of regionalized factors is possible thanks to the model of coregionalization's properties. More specifically, to account for decomposition [8], the cross covariance between the random variable $Z_i(x)$ and the lth RF of the kth spatial structure (Y_{lk}) is determined as (Goovaerts 1997):

$$Cov\{Z_i, Y_{lk}\} = a_{il}^k \, c_{a_i 0}^k \tag{14}$$

where $c_{a_i 0}^k$ is the covariance (for the kth structure) between point a_i (where the ith variable has been measured) and the location where the prediction is required, while a_{il}^k is the element of the ith row of the lth column of matrix $\mathbf{A}_k$ determined by (12). The optimal cokriging weights $\lambda_{\beta_l l}^k$ assigned to each data location for each regionalized factor are provided by the solution of the following cokriging system of equations:

$$\begin{cases} \sum_{j=1}^{p} \sum_{\beta_j=1}^{n_j} \lambda_{\beta_j l}^k \, C_{ij}\left(\mathbf{u}_{\alpha_i} - \mathbf{u}_{\beta_j}\right) + \mu_{il}^k = a_{il}^k c^k \left(\mathbf{u}_{\alpha_i} - \mathbf{u}\right) & \alpha_i = 1,......, n_i \qquad i=1,....,p \\ \sum_{\beta_i=1}^{n_i} \lambda_{\beta_i l}^{\,k} = 0 & i=1,....,p \end{cases} \tag{15}$$

where n_i and n_j are the number of sample locations for the original variable i and j, respectively, while   is the auto- (or cross-) covariance for variables i and j between locations. Finally, is the Lagrange multiplier for the ith variable and the lth regionalized factor.

The cokriging system in (15) differs from the classical cokriging system as far as the way to compute the cross covariance between primary and secondary data and in the unbiasedness constraints is concerned. More specifically, since the regionalized factors have (by definition) a zero mean, the cokriging weights have to sum to zero for the system to be overall unbiased.

2.2.7 Example: Spatial estimation of regionalized factors in the Duero

The first regionalized factor of the long-range scale of variation presents an area of high positive values in the southern part of the Duero basin. This area is characterized by a higher metal concentration for all the analyzed elements. According to the results presented in Section 2.2.3, the basin's great lithological features are responsible for enriching this area with heavy metals. Conclusively, the variability depicted in this map, representing a juxtaposition between metal-rich areas and poorer ones, is caused by the chemical composition of the geological substrate; human sources of heavy metals are not obvious in this map.

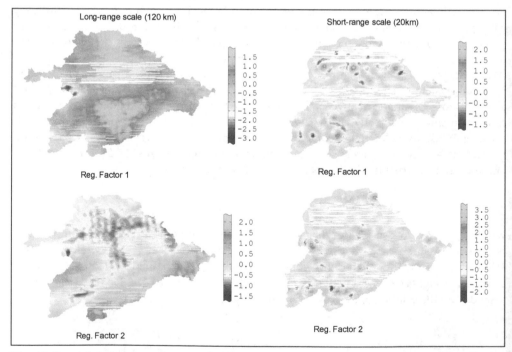

Fig. 10. Spatial estimation of regionalized factors for the Duero case study. Note that several grid points in the short-range spatial components cannot be estimated due to the limited number of data points. Thus, the study areas seem smaller in the maps of the short-range structure.

In contrast, variability in the first regionalized factor of the 20 km range scale is, according to the previous results, due to man-made enrichment of soil with Hg. Therefore, areas with higher regionalized factor values for this map depict those areas where mercury enrichment by human activities has been observed. The second regionalized factor for the same scale of variation depicts small-scale variability in all elements –except the previously reported Hg– which is due to the natural edaphic properties of soil (texture, soil organic matter, etc.) observed on small spatial scales..

3. Conclusion

Several human activities, such as agriculture, mining or industry, have dramatically increased the concentration of some heavy metals in soil causing, in some cases, severe soil pollution. Source identification of heavy metals can prove to be a great step forward in the prevention of soil pollution and rationalization of soil management practices. However, when multiple sources of heavy metals contribute to total concentration and are mixed with natural metals in soil, then the task of identifying and assigning pollutant sources remains an unresolved problem.

Fortunately, some human heavy metal sources enrich soil on different spatial scales. For instance, the impact of fertilization of crops in arable soil is on a rather local range. This is the case of Zn, Cu, Pb and Cd on the local scale of variation in the Ebro basin. Conversely, contaminant sources, such as industrial plants, disperse large quantities of heavy metals over long dispersal distances. The most obvious case was detected in both the Ebro and Duero basins for the spatial distribution of Hg. Finally lithological features, which also control the –natural- spatial distribution of heavy metals, may have a much broader range of spatial distribution than human enrichment factors. This was the case in the Ebro basin where Cd distribution on the longest-range spatial scale was found to be controlled by lithology.

A factorial kriging analysis was presented as a tool to provide some evidence about the source identification of soil heavy metals. From a statistical viewpoint, the method relies heavily on a scale-dependent decomposition of the original variables (the heavy metal concentration) in a new set of composite functions (regionalized factors). These new unobservable functions may be interpreted using the correlation circle and spatial distribution maps. Regionalized factors have the remarkable property of being scale-specific. Regionalized factors are constructed based only on the scale-specific correlations among the original variables. Thus they can be used to filter out variability on unwanted spatial scales and to reveal scale-specific common sources of heavy metals in soil.

4. Acknowledgment

We appreciate the financial assistance provided by the Spanish Ministry of Innovation through project JC2010-0109. We are also grateful to Ministerio de Ciencia e Innovacion, proyect: CGL2009-14686-C02-02 and to CAM project: P2009/AMB-1648 CARESOIL.

5. References

Alloway, B.J. (1995). Heavy metals in soils. Chapman & Hall, (Ed.). Glasgow, UK.

Alloway, B.J. & Jackson, A.P. (1991). The behaviour of heavy metals in sewage sludge-amended soils. *The Science of the Total Environment*, Vol. 100, pp. 151-176.

Boluda, R. ; Andreu, V. ; Pons, V. & Sánchez, J. (1988). Contenido de metales pesados (Cd, Co, Cr, Cu, Ni, Pb y Zn) en suelos de la comarca La Plana de Requena-Utiel (Valencia). *Anales de Edafología y Agrobiología*, Vol. 47, No. 11-12, pp. 1485-1502.

Brunet, M. F. (1986). The influence of evolution of the Pyrenees on adjacent basins. *Tectonophysics*, Vol. 129. pp. 343-354.

Castrignanò, A. ; Giugliarini L. ; Risaliti R. & Martinelli, N. (2000). Study of spatial relationships among some soil physico-chemical properties of a field in central Italy using multivariate geostatistics. *Geoderma*, Vol. 97, No 1-2, pp. 39-60

Chilés, J.P. & Delfiner, P. (1999). Geostatistics: modeling spatial uncertainty. John Wiley & Sons (Ed.), New York, 695 pp.

Chojnacka, K.; Chojnacki, A.; Górecka, H. & Górecki, H. (2005). Bioavailability of heavy metals from polluted soils to plants. *The Science of the Total Environment*, Vol. 337, pp. 175-182.

Colgan, A.; Hankard, P.K.; Spurgeon. D.J.; Svendsen, C.; Wadsworth, R.A. & Weeks, J.M. (2003). Closing the loop: A spatial analysis to link observed environmental damage to predicted heavy metal emissions. *Environmental Toxicology and Chemistry*, Vol. 22, pp. 970-976.

De Meeûs, C.; Eduljee, G. & Hutton, M. (2002). Assessment and management of risks arising from exposure to cadmium in fertilizers. *The Science of the Total Environment,*Vol. 291, pp 167-187.

De Temmerman, L.; Vanongeval, L.; Boon, W.; Hoenig, M. & Geypens, M. (2003). Heavy metal content of arable soils in northern Belgium. *Water, Air and Soil Pollution*, Vol. 148, pp. 61-76.

Dudka, S.; Piotrowska, M. & Chlopecka, A. (1994). Effect of elevated concentrations of Cd and Zn in soil on spring wheat yield and the metal contents of the plants. *Water, Air, and Soil Pollution*, Vol. 76, pp. 333-341.

Engle, M.A.; Gustin, M.S.; Lindberg, A.W. & Ariya, P.A. (2005). The influence of ozone on atmospheric emissions of gaseous elemental mercury and relative gaseous mercury from substrates. *Atmospheric Environment*, Vol. 39, pp. 7506–7517.

Errecalde, M.F.; Boluda, R.; Lagarda, M.J. & Farre, R. (1991). Indices de contaminación por metales pesados en suelos de cultivo intensivo: aplicación en la comarca de L'Horca (Valencia). *Suelo y Planta*, Vol.1, pp. 483-494.

Facchinelli, A. ; Sacchi, E. & Mallen, L. (2001). Multivariate statistical and GIS-based approach to identify heavy metal sources in soil. *Environmental Pollution*, Vol. 114, pp. 313-324.

Gómez, J.J.; Lillo, J. & Sahún, B. (2006). Naturally occurring arsenic in groundwater and identification of the geochemical sources in the Duero Cenozoic Basin, Spain. *Environmental Geology*, Vol. 50, pp. 1151–1170.

Goovaerts, P. (1992). Factorial kriging analysis: A useful tool for exploring the structure of multivariate spatial information. *Journal of Soil Science*, Vol. 43, pp. 597-619.

Goovaerts, P. (1997). Geostatistics for natural resources evaluation. Applied geostatistics series. Oxford University Press (Ed.), New York.

Goulard, M. & Voltz, M. (1992). Linear coregionalization model: tools for estimation and choice of cross variogram matrix. *Mathematical Geology*, Vol. 24, No 3, pp. 269-286.

Gzyl, J. (1999). Soil protection in Central and Eastern Europe. *Journal of Geochemical Exploration*, Vol. 66, pp. 333-337.

Houba, V.J.G.; Van Der Lee, J.J. & Novozamsky, I. (1995). Soil and plant analysis, a series of syllabi. Part 5B: In: *Soil analysis procedures, other procedures*. Dept. of Soil Science and Plant Nutrition, Agricultural Univ (Ed.), Wageningen.

Hutton, M. (1982). Cadmium in European Community, In: *MARC Report Nº 2*, MARC (Ed.), London.

Isatis (2008). Isatis software manual. Geovariances & Ecole des Mines de Paris, 579 pp.

ISO 11466. 1995. Soil quality: Extraction of trace elements soluble in agua regia, Genova, ISO.

Kashem, M.A. & Singh, B.R. (2001). Metal availability in contaminated soils: Effects of flooding andorganic matter on changes in Eh, pH and solubility of Cd, Ni and Zn. *Nutrient Cycling in Agroecosystems*, Vol. 61, pp. 247–255.

Mantovi, P. ; Bonazzi, G. ; Maestri, E. & Marmiroli, N. (2003). Accumulation of copper and zinc from liquid manure in agricultural soils and crop plants. *Plant and Soil*, Vol. 250, pp. 249-257.

Matheron, G. (1982). Pour une analyse krigeante de données régionalisées, Centre de Géostatistique, Fontainebleau.

Martin, H.W. & Kaplan, D.I. (1998). Temporal changes in cadmium, thallium, and vanadium mobility in soil and phytoavailabilty under field conditions. *Water, air and soil pollution*, Vol. 101, pp. 399–410.

Navarro, M.; López, H.; Sánchez, M. & López, M. (1993). The effect of industrial pollution on mercury levels in water, soil, and sludge in the coastal area of Motril, Southeast Spain. *Archives of Environmental Contamination and Toxicolgy*, Vol. 24, pp. 11–15.

Nicholson, F.A.; Smith, S.R.; Alloway, B.J.; Carlton-Simith, C. & Chambers, B.J. (2003). An inventory of heavy metal input to agricultural soil in England and Wales. *The Science of the Total Environment*, Vol. 311, pp. 205-219.

Nriagu, J.O. (1990). Global metal pollution. Poisoning the biosphere?. *Environment*, Vol. 32, pp. 28-33.

Olte, M.L.; Haarsma, M.S.; Broekman, R.A. & Rozema, J. (1993). Relation between heavy metal concentrations and salt marsh plants and soil. *Environmental Pollution*, Vol. 82, pp. 13-22.

Riba, O.; Reguant, S. & Villena, J. (1983). Ensayo de síntesis estratigráfica y evolutiva de la cuenca terciaria del Ebro. In: *Geología de España* , pp. 41-50, IGME Jubilar J. M. Ríos.(Ed.), Madrid, Spain.

Ross, S.M. (1994). Toxic metals in Soil-Plant Systems. John Wiley and Sons Ltd. (ed SM Ross). Chischester.

Rodríguez Martín, J.A.; López Arias, M. & Grau Corbí, J.M. (2006). Heavy metal contents in agricultural topsoils in the Ebro basin (Spain). Application of multivariate geostatistical methods to study spatial variations. *Environmental Pollution*, Vol. 144, pp. 1001–1012.

Rodríguez Martín, J.A.; López Arias, M. & Grau Corbí, J.M. (2009). Metales pesados, materia organica y otros parametros de los suelos agricolas y de pastos de España. MARM-INIA (Ed.). ISBN 978-84-491-0980-5. Madrid. Spain.

Rodríguez, J.A.; Nanos, N.; Grau, J.M.; Gil, L. & López-Arias, M. (2008). Multiscale analysis of heavy metal contents in Spanish agricultural topsoils. *Chemosphere*, Vol. 70, pp. 1085–1096.

Roure, F.; Chokroune, P. & Deramond, J. (1989). ECORS deep seismic data and balanced croos-sections, geometric constraints to trace the evolution of the Pyrenees. *Tectonics*, No 8, pp. 41-50.

Sanchidrian, J.R. & Mariño, M., (1980). Estudio de la contaminación de suelos y plantas por metales pesados en los entornos de las autopistas que confluyen en Madrid. II contaminación de suelos. *Anales de Edafología y Agrobiología*, Vol. 39, pp. 2101-2115.

Saporta, A. (1990). Probabilités, analyse des données et statistique. Technip, Paris.

Simón, J. L. & Soriano, M. A. (1986). Diapiric deformations in the Quaternary deposits of the central Ebro basin, Spain. *Geological Magazine*, Vol. 123, pp. 45-57.

Simon-Gomez, J. L. (1989). Late Cenonzoic stress field and fracturing in the Iberian Chain and Ebro Basin (Spain). *Journal of Structural Geology*, Vol. 11, No3. pp. 285-294.

Söderström, M. (1998). Modelling local heavy metal distribution: a study of chromium in soil and wheat at ferrochrome smelter in south-western Sweden. *Acta Agriculture Scandinavica*, Vol. 48, pp. 2-10.

Tejero, R.; González-Casado, J.J.; Gómez-Ortiz, D. & Sánchez-Serrano, F. (2006). Insights into the "tectonic topography" of the present-day landscape of the central Iberian Peninsula (Spain). *Geomorphology*, Vol. 76, pp. 280– 294.

Tiller, K.G. (1989). Heavy metals in soils and their environmental significance. *Advances in soil science*, Vol. 9, pp. 113-142.

Wackernagel, H. (1998). Multivariate geostatistics: an introduction with applications. Springer, 285 pp.

Weber, J. & Karczewska, A. (2004). Biogeochemical processes and the role of heavy metals in the soil environment. *Geoderma*, Vol. 122, pp. 105-107.

3

Potential and Geochemical Characteristics of Geothermal Resources in Eastern Macedonia

Orce Spasovski

University "Goce Delcev" Stip, Faculty of Natural and Technical Sciences
Macedonia

1. Introduction

Geothermal explorations in the Republic of Macedonia were intensified in the 70's, during the first effects from the energetic crisis. As a result of those explorations, there were established over 50 springs with mineral and thermo-mineral water, with maximum potential of over 1400 l/s and evidenced reserves as deposit for exploitation of around 1000 l/s, with temperature higher than the mean season swings for this part of the Earth in the range 20-79 °C, accumulated quantities of geothermal power. Geologically and hydro-geologically spoken, these geothermal resources are mainly located in eastern and southeastern Macedonia, in the Bregalnica-Strumica region composed by 23 municipalities. This region, in its geological past has undergone many big tectonic changes and its composition is formed by almost every type of stones, including the youngest and oldest formations, forming hydro geothermal systems, that currently are the only ones worthwhile to be explored and exploited. The most important hydro geothermal systems in this area are the tertiary basins of Kochani valley, Kezhovica Spa near Shtip, and the Kratovo-Zletovo volcano region, exactly the micro regions included in the REDEM region as the territory from the project target group. In order to valorize this geothermal potential in eastern and southeastern Macedonia certain pre-conditions should be created for using these resources in Bregalnica-Strumica region. This means that serious project documentation should be prepared followed by the targets that should be accomplished and tasks for faster development in order to use this important geothermal potential in this region, but at the same time following the positive legislation for issuing a concession for carrying out Detailed Geological Explorations (DGEs) and exploitation of the geothermal energy in the Republic of Macedonia (RM), according to the Law on mineral raw materials (Official Gazette of the RM, no. 18/99 and 29/02). With the project "Creation of pre-conditions for utilization of the geothermal potential in the eastern and south-eastern region", the basic information about the geothermal resources in the RM will be presented in brief. In addition, the project will present details of the potentials of this region showing the energetic value of the geothermal resources in this region, currently available capacities, technical capabilities and deposit characteristics.

Simple definition of the geothermal energy is that it is a quantum of calories (the thermal energy in warm water) that the Earth holds in its inside part. According to the location where the geothermal energy is accumulated, in which stratum of the Earth's surface, it can be divided into: hydro geothermal, litho geothermal, magma geothermal and pneumonic geothermal energy. The geothermal energy can be accumulated in the fluids, in the solid rocks, in magma and in gases. Definition of geothermal resource is that it is a part from the Earth's inside heat (higher than the mean annual temperature) that can be exploited, using its economic effect but following the legislation, nowadays or in the near future. Hydro geothermal resources are part of the geothermal resources and presents the part of the Earth's stratum in which apart from the conductive way of transfer of the geothermal energy through rocks as isolators, is also transferred by a convective way among the rocks and theirs joints that serve as reservoirs, which calories can be rationally used as energy, i.e. exploited. According to the way this geothermal energy is being transferred near the Earth's surface, the hydro geothermal resources are divided in conductive and convective systems.

The group of conductive systems gathers the systems with dominant conductive way of transfer of the geothermal energy. In this group we find the tertiary basins of Ovce Pole, Tikvesh, Delchevo-Pehchevo region, Skopje valley, Strumica valley, Gevgelija valley, Polog and Pelagonija valley (Arsovski and Stojanov, 1995). The group of convective systems is developed in crags, capillary crags and with capillary porosity. Major characteristics of all convective systems are: high hydraulic gradients and high pressures due to the big difference in the heights levels of zones of influx (where the underground water enters the reservoir) and the zones of outflow, where the underground water leaves the pool (Serafimovski, 2001). Before we analyze the geo-thermal potentials of eastern and southeastern Macedonia, we will present the main characteristics of the hydro geothermal resources in the RM.

2. Research methodology

Since the study subject have been terrains with complex geological evolution, an application of complex methodology of field and modern analytical methods, was required which should provide more exact data for realistic interpretation of established goals.

Efforts were made for the first time in this researched area and one large set of geological and non geological data were collected during a longer period of time and studies were conducted along different criteria, their analysis and synthesis by same criteria.

During these studies geological and non-geological methods were used. In regards of geological methods were applied: analysis and synthesis of previous geological studies with special review of tectonic studies, geotectonic concepts and distribution of formations of carbonaceous composition.

These non-geological methods were applied: analysis and synthesis of data from chemical analyses of thermal waters with special attention given to sampling of specimens and applied analytical methods, analysis and synthesis of chemical studies and temperature measurements in realized drill holes.

3. Geothermal resources in Macedonia

Geothermal energy is the warmth that the Earth holds in its inside part. This energy comes from the inner nature of our planet, the physical processes that occur underground. Getting to and using this warmth is conditioned by a certain carrier - that transports the energy like warmth near the Earth's surface, at considerable depth. This carrier usually is some fluid.

Hydro geothermal energy and resources nowadays are the only ones of the commercially approved sources of energy kept in hot/warm water or steam, accumulated in rock's joints or in porous sedimentary structures at depths going from few hundreds to 4000 m underground. Depending on the temperature and the dominant phase of the fluid, the hydro geothermal resources generally are categorized in 3 groups: low temperature water, mainly used for central heating in homes, apartments, offices, supporting agricultural production in greenhouses (hothouses), and few industrial processes; medium temperature water (<140°C), used in some industrial processes and production of electric energy (using binary cycles-plants that use Freon, known as ORC turbines); and high temperature water or steam (140-350°C) is used for electric energy production in turbine plants (of which 2/3 work in the moderate area 150-200°C).

3.1 Geological characteristics and tectonic placement of the RM

Macedonia is situated in the middle of the Balkans peninsula and has 25.713 km². The geotectonic position is determined by its location. It belongs to the Alps - Caucasus - Himalaya geosyncline belt, and the history in creating of the terrain is tightly connected with the former geosynclinals Tetis and the Alpine orogeny, also with the primordial position of the lithosphere. Generally spoken, over the components of the Alps orogen, the territory of RM belongs on two tectonic systems: the western part including the Vardar valley belongs to the Dinarides (Helenides) whereas the eastern Macedonian mountainous terrain and valley's depressions are segment of the Serbian-Macedonian massif. Additionally, along the Macedonian-Bulgarian borderline, a small separated zone belongs to the Carpatian-Balkanides, the Kraishtide zone (Dimitrievic, 1974).

According to this geotectonic reorganization, the main part of the Macedonian territory that spreads west from the Dojran-Strumica-Zletovo-Kumanovo line is divided in four (4) geotectonic units: Vardar zone, Pelagonian horst-anticlinorum, western Macedonian zone, and the Korab zone better known as Cukali-Krasta zone Fig. 1.

To the East from the mentioned line lies the Macedonian massif of mountains which is connected with the Rodopi Mountains in Bulgaria through the Ograzhden Mauntain. In the borderline zone with Bulgaria, to the north of Berovo-Delchevo, as a wedge in the old Rodopi Mountains there are elements from the Carpatian-Balkanides massif separated in the Struma zone (better known as Kraishtide zone, after I. Bonchev).Every separate unit is one structural entity, characterized with its separate geological development including the specific processes of tectonic deformations and manifestation of magma differentiation.

The terrain that forms the territory of the RM in its geological past has undergone big tectonic changes and it is represented by almost every type of rocks, from the oldest to the youngest, starting with Precambrian metamorphosis rocks, with high cristalinity and than the youngest neogenesis and quarter sedimentary complexes. Also, there are found vast surfaces with eruptive rocky masses, from ultra basal to extremely acid and alkaline magma rocks. As a result of all above mentioned, we have found forms with highly specific geological, hydro geological and geomorphologic characteristics. Because of the heterogeneous geologic-lithologic composition and the tectonic past of the terrain as well as from the different geomorphologic and climate characteristics, there are found different types of underground water. In the mountainous part of the terrain there are more dispersed springs usually found, as opposed to the numerous valleys, there are also springs that are more compactly settled, some with free surface and other under pressure (artesian wells-springs), while in the seismic active zones we find the thermal and mineral waters, i.e. the hydro geothermal resources.

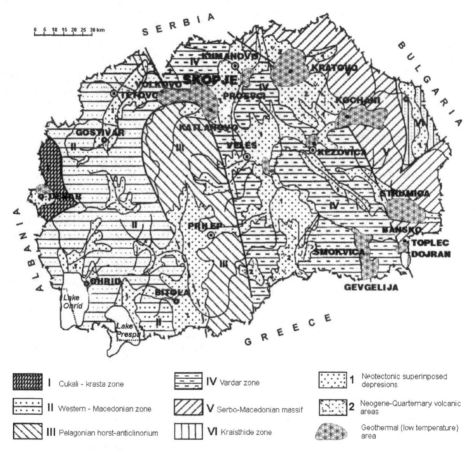

Fig. 1. Main geothermal areas in the Republic of Macedonia and regional tectonic setting (Arsovski, 1997)

3.2 Geothermal characteristics

According to the way of transmission of the geo-thermal energy, hydro-geo-thermal resources are divided in two groups: convective and conductive hydro-geo-thermal systems. The group of conductive systems gathers the systems with dominant conductive way of transfer of the geothermal energy. In this group in the RM we find the tertiary basins of Ovce Pole, Tikvesh, Delchevo-Pehchevo basin and the basins of Skopje valley, Strumica valley, Gevgelija valley, Polog and Pelagonija valley. In the group of convective hydro geothermal systems a convective way of transfer of the geothermal energy dominates. The most important convective hydro geothermal systems in the territory of the RM are: Skopje valley, Kocani valley, Strumica valley, Gevgelija valley, Kezhovica near Shtip, Toplic-Topli dol on the Kozuf Mountain, Toplec near Dojran, Proevci near Kumanovo, Strnovec, Zdravevci at the river Povishnica near Kratovo, Sabotna voda near Veles, as well as the systems in the western Macedonia - Kosovrasti and Debar spa, and Banice at the river Pena near Tetovo.

In the following text we will present the data and the parameters on the most important hydro geothermal systems in the eastern and south-eastern part of Macedonia. Location of the main geothermal fields and their systems are shown on Fig. 2.

On the Macedonian territory there are rocks from different ages, starting from Precambrian to Quarter, presented by almost all litho logic types. The oldest, Precambrian rocks are composed of gneiss, micaschists, marlstone and orthometamorphides, the Paleozoic rocks are mostly green schists and Mesozoic are presented by marlstone, limestone, acid-, alkali- and ultra alkali magma stones. Tertiary sediments are composed of flysch and lake sediments, sandstones, limestones, claystones and sands. What stands to the structural relations the territory can be divided in six geotectonic units: Chukali Krasta zone, west Macedonian zone, Pelagonian horst anticlinorum, Vardar zone, Serbian-Macedonian massif and Kraishtida zone. The tectonic position is based upon the terrain itself and the geological data without using geo-thermal hypothesis (Arsovski, 1997). The first four tectonic units are part of the Dinarids, the Serbian-Macedonian massif is part of Rodopi and Kraischtida zone is part of Carpatian-Balkanidi zone.

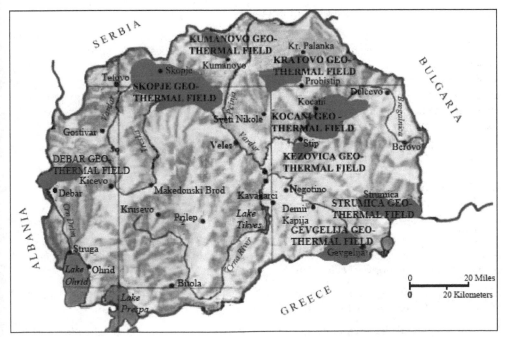

Fig. 2. Main geothermal fields in Macedonia (Popovski & Micevski, 2005)

The territory of the RM as a whole is part of the Alps-Himalayan zone, the Alps sub zone that is still volcanic active. This chain starts in Hungary, runs across Serbia, Macedonia and Northern Greece and all the way to Turkey. Few geothermal regions, including the Macedonian region that is connected with Vardar tectonic unit are set aside. This region shows positive geothermal anomalies. The hydro geothermal systems currently are the only one worthy to be explored and they are exploited. There are 18 geothermal fields known in the country, with over 50 thermal springs, boreholes and hot water wells. The maximal capacity is 1000 l/sec water flow with temperatures from 20-79°C. Thermal water is mostly hydrocarbon according to the dominant anion and equally mixed with Na, Ca and Mg. The concentration of the dissolved minerals stands from 0, 5-3, 7 g/l. All thermal waters in Macedonia are with meteoric origin. The thermal spring is a part from the regional thermal flow, which in the Vardar zone 100-120 mW/m2, while in other part of the country is 60-80 mW/m2, with 32-35 km thick stratum. The following text will present the details on the data and the parameters for the most important convective hydro geothermal systems in eastern and south-eastern Macedonia, a territory composed of 23 municipalities from Bregalnica-Strumica region.

4. Main geothermal fields in eastern Macedonia that are currently exploited

This presentation will show basic data about the energetic resources in eastern and southeastern Macedonia, current capacities in Macedonia, the technical capability and perspective locations in this part of the country, as well as the methodologies for their valorization from the geothermal aspect.

4.1 Hydro geothermal system of Kochani valley

The geothermal locality "Podlog"-Banja-Kochani is not only important deposit of geo-thermal water in Macedonia, but also in the world (verified with the studies of the American enterprise GeothermaEx and the Austrian consortium ARGE GTM), having in mind the balanced reserves from 157x106 m3 GTW, with exclusively good chemical characteristic, and mean temperature of 75°C. This great potential of the deposit from the aspect of energy, especially the characteristic restorable reserves, has offered the main prerequisite for preparation of the concept, project and implementation of the exploitation system GTW "Geoterma".

The geothermal system "Geoterma" with its installed capacity 300 l/sec exploits and distributes geothermal water to the end users:

1. Heating the Agro-complex - production in greenhouses
2. Low-temperature procedures
3. Central heating of public and administrative buildings 4. Recreation centers and balneology

Basic data for the geothermal deposit Podlog-Banja (geography, morphology, geology, petrology, hydro-geology and chemistry).

Kochani geothermal region i.e. deposit Podlog-Banja is situated in the north-eastern part of Macedonia between the 41° 40' and 42° 00' NGW and 22° 00' and 22° 30' EGL.

Kochani valley with its surroundings is generally spread in East/ West direction and has about 400 km2, medium altitude of 330 m. By its morphology the valley is an elongated field in the medial part of the Bregalnica river basin. The economic, cultural and political center is the city of Kochani (30.000 inhabitants).

The deposit itself in regional-geological sense belongs to the zone of higher thermal flow that stretches from Turkey, across northern Greece, eastern Macedonia all the way to the Panonian basin. In a tectonic sense the region is a complicated orogene area that belongs to the two tectonic units: Serbian-Macedonian basin and the Vardar zone with highly expressed volcanic activity of the Kratovo-Zletovo volcanic area Fig. 3.

The deposit from the point of view of supplementation is an infiltration type.

Based upon the completely done interdisciplinary researches, it is verified as maybe one of the world largest non-magma deposit of geothermal water, with overall balanced static reserves of approximately 150.000.000 m³ over 70°C. The chemistry of the water is Na-bicarbonate water with pH=7, none corrosive enclosed systems, that gives its possibility for various applications. The presence of some elements (Se, F and etc.) in controlled referent limits, gives the water special quality and it can be used as a bottled drinking water (Table 1).

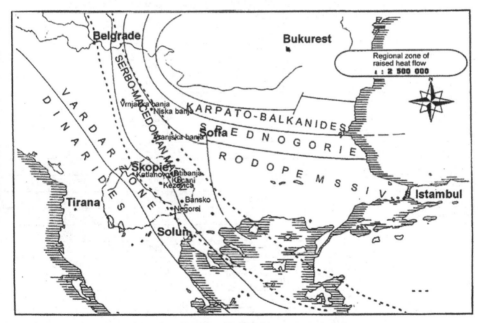

Fig. 3. Regional zone with increased thermal flow (Bonchev, 1974)

In order to take a look on the activities that practically have verified the deposit of geothermal water in "Podlog-Banja" and established the "Geoterma" system, we will present a review, chronology and results organized in three temporal stages (Naunov, 2003).

4.1.1 First stage

The first stage is typically explorative and is located at the very beginning, in the year 1973 and all the way to the year 1984. This period, organized by the Council of the Municipality

	Name of the chemical substance	Average value mg/l	MDK, for drinking water mg/l
1	Color, Co/Pt	< 5	100
2	starring, NTU	0.7	< 0.6
3	dry remrant on – 105 o S, mg/l	599.6	-
4	dry remrant on – 180 o S, mg/l	620.0	-
5	dry remrant on	493.5	< 500
6	pH – value	6.99	6.8 – 8.5
7	p – value, ml 0.1 N HCl/l	0	-
8	m – value, ml 0.1 N HCl/l	86.4	-
9	Bicarbonate ions, [HCO_3], mg/l	527.4	-
10	total solidity, [VT], o D	9.27	-
11	Carbonate solidity, [KT], o D	23.02	-
12	Calcium solidity, [Ca T], o D	6.08	-
13	Magnesium solidity, [Mg T], o D	2.90	-
14	Chlorides, [Cl], mg/l	17.9	25.0
15	Sulphates, [SO_4], mg/l	50.38	25.0
16	Calcium, [Ca^{2+}], mg/l	48.32	100.0
17	Magnesium, [Mg^{2+}], mg/l	11.71	30.0
18	Total Fe, [Fevk], mg/l	0.14	0.05
19	Manganese, [Mn^{2+}], mg/l	0.019	0.02
20	Sodium, [Na^+], mg/l	142.4	20.0
21	Potassium, [K^+], mg/l	18	10.0
22	Silicon Dioxide, [SiO_2], mg/l	44.2	-
23	Free CO_2, mg/l	109.8	-
24	Aggressive CO_2, mg/l	20.7	-
25	Conductivity	775	to 300
26	Nitrites, [NO_2], mg/l	0.0034	-
27	Nitrates, [NO_2], mg/l	0.29	5.0
28	Ammonium, [NH_4^+], mg/l	0.305	0.01
29	Phosphate, [P_2O_5], mg/l	2.29	0.03
30	Solubler oxygen, [O_2], mg/l	5.16	-
31	Cyanides, [CN], mg/l	0.0	-
32	Sulphides, [S], mg/l	0.0	-
33	Lead, [Pbvk], mg/l	< 0.01	0.05
34	Cadmium, [Cd^{2+}], mg/l	< 0.002	0.005
35	Chromium, [Crvk], mg/l	< 0.005	0.05
36	Copper, [Cu^{2+}], mg/l	< 0.005	0.1
37	Zinc, [Zn^{2+}], mg/l	0.14	0.1
38	Nickel, [Ni^{3+}], mg/l	< 0.005	0.01
39	Consumption of $KMnO_4$	2.13	to 5

Table 1. Summary values of the results from the chemical analyses of the geothermal waters on the locality Podlog – Banja, Kocani (Naunov, 2003)

of Kochani - Study and research office and the Institute on explorative works in mines of RM (Republic of Macedonia), realized by the Geologic institute Skopje and Ljubljana (Slovenia) and GeothermEx (USA), the following researches have been done:

Detailed geologic and hydro-geologic explorations

Detailed structural-petrologic works

Geophysical works (deep geophysics, gravimetric, seismic and shallow geo-thermometric)

Micro location of the exploration boreholes; Based upon the synthesis of all gathered information from the exploration, officially elaborated the following results: the geothermal area in the Kochani valley is being fully characterized in all segments (geologic, petrologic, geophysical, structural and hydro-geological characteristics).

4.1.2 Second stage

Activities in the second stage were mainly organized around promotion of the results and the detailed explorative works from the first stage at the location Podlog-Banja, all with a role of creating an objective ground for establishing a re-injection system. The period of these activities is after 1984 and is characterized with organizing new personnel, at the beginning like an investment group within PCE "Vodovod"-Kochani, and later transformed in an independent experts group of the geothermal system "Geoterma". Since 1984 till present it is the main initiator and animator of the activities concerning the system.

During this period a lot of domestic and foreign experts, also exploration institutions took part and successfully finished the following tasks:

-    located and drilled exploiting wells EW MP-1, EW2, EW3 as well as re-injecting borehole P-10, and then in 1999, well EW-4;
-    for utilization of the capacities of the deposit 'Podlog-Banja', the concept was prepared, feasibility study elaborated and the project prepared of the exploitation system GTW "Geoterma" (Fig. 4.)
-    The system "Geoterma" was built and activated as an exploitation-distribution system; the basic "Geoterma" technical-technological parts are:
-    Exploitation wells for geothermal water (GTW)
-    Pipeline
-    Pumping and distributing unit
-    Distribution pipeline for GTW

At this phase of development, the system has installed capacity of 300 l/sec, and practically an annual average of 1.400.000 m³ GTW pumped and distributed with mean temperature of 75°C.

The main usage of the geothermal water is for the needs of:

-    agricultural production in greenhouses, around 18 ha
-    low-temperature technological processes
-    Central heating of public and administrative buildings in the downtown Kochani

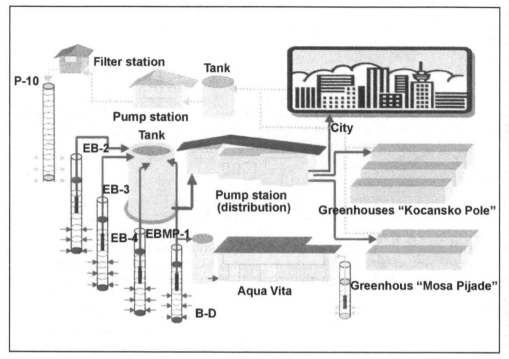

Fig. 4. General scheme of the System for exploitation, distribution and reinjection "Geoterma" – Kochani (Naunov, 2003)

Partial realization of the investment-technical conception of the "Geoterma" system as well as the findings and the results gained during the exploitation phase in the period of 1985-89 with registration of the piesometric level of the basin, clearly and without doubts has shown an urgent need of building of the re-injecting phase of the system (Fig. 5).

For this purpose a ^{3}H-tritium traced detailed explorative researches have been done and at the same time the unlimited possibility of applying of the re-injection was proven. Afterwards, upon these results and the created project solutions, the basic part of the re-injecting system was realized as an inseparable i.e. complementary part of the whole exploitation system and practically encircled the system "Geoterma".

The re-injection system in this stage of equipment has the main role to cap the used quantities of GTW from the agro complexes (at this stage only from AC "Mosha Pijade"-Podlog) and through the re-injection pump system to inject it back in the re-injection borehole P-10 in the village of Banja. But, besides the big expectations from this re-injecting system, from well known reasons-transformation of 2-valence Fe in to 3-valence Fe in the GTW and creation of voluminous sediment threatening to close the re-injecting borehole-it didn't work well with full capacity.

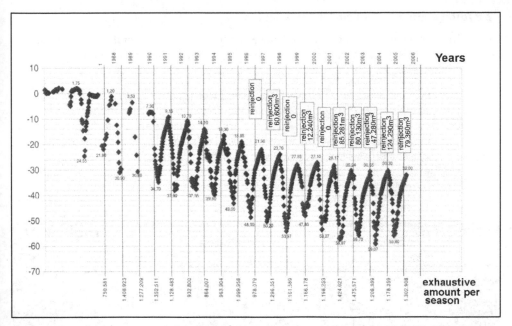

Fig. 5. Chart of the piesometric level of the basin in the period from 01.01.1987 -19.10.2006

4.1.3 Third stage

This stage of the activities: 1995-2003, treats mainly the problems that originated from our experience in exploitation of GTW, as well as problems with development program for the system "Geoterma". Organized as it is with very few innovations and participation of renowned experts and institutions on this field, in this stage we should emphasize the following activities:

- intensification of the researches for solving the problem with the re-injecting system;
- technical-technological equipping of the system;
- researches and preparations for implementation, and the implementation of the project-bottling of GTW;
- encircling of the central heating system for the town of Kochani
- continuity in the research activities for defining the mechanisms of supplementation of the deposit, and completely defining of the one (shown at the prognostic pattern, (Fig. 6);

The other effects are clearly expressed in few segments, above all in using the GTW for bottling, with accent on its outstanding quality and the special wellness characteristics positive for the health of the consumers, as well as for balneology-recreation purposes;

In the scientific-research segment it is a special contribution as applied explorative model and its results;

In the administrative-legislative segment - also as an avant-garde object-project, that helped the development of geothermal science and utilization of geothermal waters in Macedonia.

4.2 Hydro geothermal system Istibanja-Vinica

The hydro geothermal system Istibanja-Vinica, with its thermo-mineral springs is directly settled in the immediate surroundings of the village Istibanja (Vinica), along the flow of river Bregalnica (Fig. 7).

The geothermal field Istibanja - Vinica from geological point of view lies on the contact between two tectonic units (the Serbian-Macedonian massif on the East and the Vardar zone on the West) at the eastern periphery of the Kochani valley, very close to Istibanja and Vinica. The microlocation of the exploitation wells 1-3,1-4 and 1-5 is between the regional motorway Kochani-Istibanja-Delchevo and the river Bregalnica, drilled in the river terrace along the river bed.

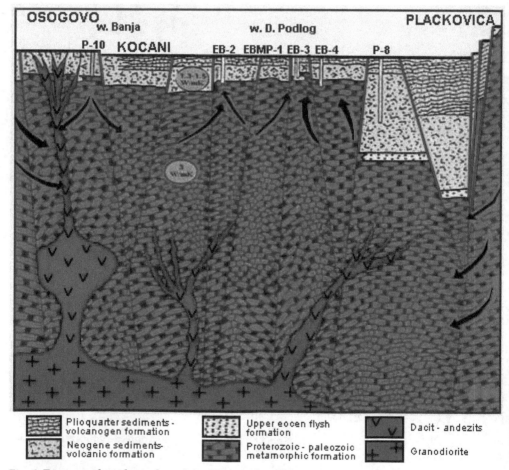

Fig. 6. Forecasted geological model of the geothermal system of the Kochani valley (Naunov, 2003)

Fig. 7. Geothermal field Istibanja-Vinica

As separated hydro geothermal system from the Kochani geothermal field (Podlog-Banja system) is micro-located on a terrain with rocky masses as part of the Serbian-Macedonian geotectonic unit.

The Serbian-Macedonian massif in the area of Istibanja and Vinica is built up of old Precambrian and Riphean - Cambrian rocks. The former, Precambrian rocks are presented by double-mica striped gneisses (Gmb), a variant of micaschists and amphibolites (Kovacevic et al 1973; .Rakicevik et al, 1973).

While analyzing the neotectonic fault zone (shear) structures, gathered as information from the satellite recordings and the geotectonic map of the RM (Arsovski , 1987) in proportional ratio 1:200.000, we can clearly match the difference between the Vardar zone on one side (western part) and the Serbian-Macedonian massif on the other (eastern part). Many ring like structures with different dimensions are typical for both systems in this region (Podlog, Banja and Istibanja). The neotectonic structural forms in the Serbian-Macedonian massif, where the Istibanja hydro geothermal system belongs are different than the tectonic structures in the Vardar zone. Also, it should be mentioned that the Vardar zone belt has deep shear structures with pretty much expressed seismic activity, contrary to the relatively stable Serbian-Macedonian massif.

The deposit-reservoir of the hydro geothermal system in the geothermal field of Istibanja-Vinica is composed of ruptured Paleozoic gneisses and granites, as shown at the prognostic model on the figure. The supplementation zone for this system is formed by the gneisses, micaschists and schist's from the Golak downhill- Krshla in the eastern part of the valley as well as from the north-eastern parts of Osogovo Mountains (Fig. 8).

The outflow zone of the system is presented with contemporary geo-thermal occurrences as natural springs in the very surrounding of Istibanja and the exploitation wells I-3, I-4 and I-5 in the very surrounding of the river Bregalnica, to the south of the village Istibanja, Vinica. According to the type of hydro-geo-thermal system, it is found in the group of semi-closed systems in ruptured (jointed) gneisses and granites from the Paleozoic age where the outflow zones are located in the shears.

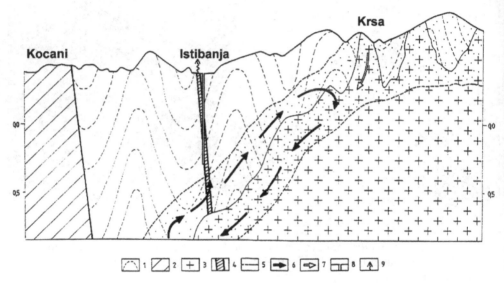

1.Pt-gneisses, 2. Pz-shall, 3. granite, 4. fault zone, 5. boundary of the reservoir hydrogeothermal, 6. superior direction of movement of thermal water, 7. superior direction of movement of cold water, 8. constructed boreholes, 9. thermal source.

Fig. 8. Forecast model of the geo-thermal field Istibanja – Vinica

The GTW from this hydro geothermal system is used in the greenhouse complex (6 ha), that is warmed using combination of geothermal water and crude oil. Supported by the grants from the Austrian government in the period after the year 2000, the system was reconstructed and prepared for production and export. The geothermal system Istibanja is using geothermal water from three exploitation wells with total capacity of 56 l/sec and water temperature of 67°C, connected with pipelines long 3,25 km. The actual available capacity that reaches the greenhouses (agricultural complex) is 50 l/sec with water temperatures of 61 °C.

4.3 Hydro geothermal system Kezhovica - Shtip

On the right shore of river Bregalnica, approximately 2 km to the southwest from the center of Shtip, on the exit of Novo Selo, lays the geothermal system Kezhovica – Ldzhi (Fig. 9).

Fig. 9. Geothermal spring Ldzhi, near Shtip

Hydro geothermal system Kezhovica - Ldzhi is near of Kezhovica spa at Novo Selo - Shtip. The reservoir of this hydro-geo-thermal system is built on Jurassic joint granites (Fig. 10). Big part of these granites is covered with tertiary sediments of Ovchepole and Lakavichki basin (Fig. 11).

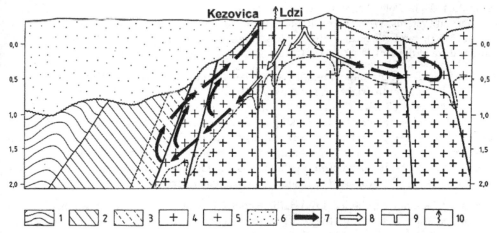

1.shall, 2. gneisses, 3. reservoir of gneisses, 4. massive granite, 5. cracked granite, 6. tertiary sediments, 7. superior direction of movement of thermal water, 8. superior direction of movement of cold water, 9. constructed boreholes, 10. thermal source.

Fig. 10. Forecasted model of the Kezhovica - Ldzi geothermal system

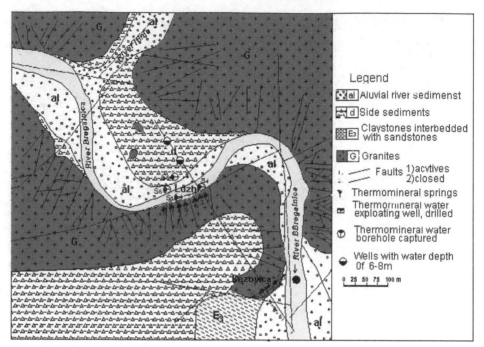

Fig. 11. Geological map of the hydro geothermal system of Kezovica Spa

Kezhovica spa today uses water from two shallow wells with capacity of 4,5 l/sec with temperature of 63°C. The supplementation of the system comes from the granite, south from Shtip. The chemic ingredient of the water shows supplementation also from the filch sediments Ovche Pole. According to the type, this system belongs to the half opened hydro geothermal systems. With chemical analysis of the water in this system is from Na-CI type (Alekin,1953), and the calculate geothermometers are applying that the forecasted temperature of the tank is 100 - 115 °C.

The site of these geothermal waters is located between the hills Isar and Kumlako from the left and Marite from the right side of the river Bregalnica. In Kezovica, 30 years ago there w only one drilled well which was used as capping in the old Shtip spa, and by the Ldzhi, as nc capped thermo mineral water leaking in few places directly on the surface. With the research made in 1953 and with information gained, on the location of the old spa in the next few years new spa has been built, with modern areas and pools and also modern rehabilitation center has been built. Kezhovica as a spa is known since the Ottoman's Empire in Macedonia, when it was used as healing thermal water, too.

Contrary to Kezhovica, the locality Ldzhi, except the relatively shallow researched drills E the formed pipelines around these drills, until these days nothing more has been done researching of this hydro geothermal system (Kekic and Mitev, 1976).

Kezhovica Spa and the thermo-mineral system Ldzhi, although are well known as import hydro geothermal phenomena especially with their great radioactivity and healing characteristic for their exploration and valorization of their geothermal fluid, nothing particular has been do Thermo-mineral waters in Kezhovica Spa are not overflowing, but today they are used by shallow wells with capacity of 4,5 l/s and water temperature of 57 - 63°C.

In Ldzhi locality there are more thermo-mineral waters in form of springs. Their maxi capacity is minimal and around 0,03-1,0 l/s. Captivated spring that exists today is with capacity 0,5-1,0 l/s. All these springs are with the same origin and are hydraulically connected, and t temperature varies from 28 - 59°C. The water in Ldzhi Spa has variable maximal capacity depends on the water in the river Bregalnica. It has the characteristic that with the raising of level of the water in the river Bregalnica, beside the capacity of the thermo-mineral water in L springs, there is uprising of the temperature of this water. That is explained with deep infiltrate of the river flow underground this terrain, through existing joints where the same has capability, besides the emerging of the capacity of the thermo-mineral water, to heat to a hi| temperature.

According to the results from ht chemical analysis of the thermo-mineral water in Kezhovica Spa and Ldzhi springs, it can be said that they are CI - Na waters. Total mineralization of the water in the capping of Kezhovica Spa is 1450 mg/l, and at Ldzhi springs varies from 1310,4 -1507,9 mg/l. Kezhovica Spa by its radioactivity belongs to one of the most radioactive thermo-mineral waters not only in Macedonia, but also in the wider region. The radioactivity in the capping Kezhovica is 42,82 Mach units, and the water in Ldzhi springs is 11,57 Mach units. This bigger radioactivity of the water in Kezhovica Spa related to the radioactivity of the water from Ldzhi springs, should be taken in consideration of the tectonic or seismic-tectonic activity of the main stratum that passes near Kezhovica, which is the main seismic dislocation that spreads from Strumica on the South through Kezhovica and Sveti Nikole all the way to Kumanovo (Kotevski, 1977).

During the years 1953 and 1964, there were six shallow explorative holes drilled (C-1 to C-6), and a large one (D-7). Their maximal capacity in terms of exploitation done in that period was 20 l/s. During the year 1983, nearby Ldzhi springs, on the right side of river Bregalnica, there was another research hole drilled, B-4 with maximum capacity of 30 l/s and with the temperature of the water 32°C. The temperature now is lowered because the water is a mixture of the surface Bregalnica water and hot water from the deep springs of cracked granites which are the tank (reservoirs) of this geothermal system, nearby Shtip.

The healing power of the water from Kezhovica Spa is in its high temperature, radioactivity and the chemical structure.

Total amount of the balanced reserves of the hydro geothermal system Kezhovica-Ldzhi are shown on Table 2.

Well number	Location	Capacity Q (l/s)	Water temperature t(°C)
S – 53 g.	Kezhovica Spa	2,7	62
B – 1/76	Kezhovica Spa	1,3	58
Total (1)	**Kezhovica Spa**	**4,0**	**58 - 62°C**
Well B – 2/97	Ldzhi	11,5	53
Well B – 3/97	Ldzhi	6,5	60
Total (2)	**Ldzhi**	**17,5**	**53 - 60°C**
Total (1) + (2)	**Kezhovica Spa + Ldzhi**	**21,5**	**53 - 62°C**

Table 2. Balance reserves of the Kozovica-Ldzhi systems

5. Utilization of the geothermal energy in the spa complexes

Macedonia although a small country is very rich in thermal and thermo-mineral water, that are currently used in the thermal spas and medical recreation centers. The geothermal energy of these geothermal and thermo-mineral waters is used in eight thermal spas, and four are located in the eastern and southeastern part of Macedonia. The thermo-mineral water in these complexes has various physical-chemical characteristics and temperatures, depending on the site and localization of the geothermal field where the spa complex is located. The water temperatures in the spas in Macedonia, starting with the highest are: Spa Bansko - 73°C, Kezhovica Spa - 54°C, Spa Negorci 38°C etc. With regards to the quantities of available flow of geothermal water in the spas in eastern and southeastern Macedonia: in Spa Bansko - 35 l/s, Kezhovica Spa - 5,4 l/s, Spa Negorci 1,8 l/s, etc. The utilization of the geothermal energy in the spa complexes in eastern and southeastern Macedonia and their thermal power are presented in Table 3.

Nr.	Location	Flow Q (kg/s)	Temp. (°C)	Thermal power for this temp.[1] 15°C (MW$_t$)	Annual usage[2] (TJ/year)
1	Spa Bansko	35	70	8,05	253.9
2	Kezhovica Spa	5,4	54	0.88	27,7
3	Negorci Spa	1,8	38	0.17	5,46
4	Kocani Spa	30	63	6,02	189,9
I	TOTAL	72,2		15,12	476,96

[1] Thermal power (MW$_t$) = Max flow Q (kg/s) h inflow T(°C) - outflow T(°C) x 0,004184 (MW=10^{6}W)
[2] Annual usage (TJ/year) = Average spend Q (kg/I) x inflow T(°C) - outflow T (°C) x 0,1319 (TJ/year)

Table 3. The utilization of the geothermal energy in the spa complexes in eastern and southeastern Macedonia and their thermal power

The planned reconstructions of the heating installations that use the geothermal water in the spa and recreational complexes in eastern and south-eastern Macedonia are still not implemented because of unsolved property issues. However, one can't expect positive developments unless the ownership over the public spa complexes is defined.

6. Comparison of the geothermal potential resources with the other sources of energy and the geothermal

In order to analyze the economic aspects of the exploitation of the geothermal energy as a power source, a unique and adequate access towards forming of the prices of every separate kind of energy source is needed, which are produced and distributed from the energetic systems, as well as for the market price of the oil at the moment when the analysis are being done. Macedonia is a part of the group of developing countries. After the independence in 1991, it entered the period of "transition" and change of the capital from public into private that also influenced the sector of industrial production and energy.

Macedonia as main power source is using domestic coal and hydro energy from power plants and imported oil. The utilization of the solar energy is insignificantly small related to the needs and the possibilities, and the utilization of the natural gas is still in a phase of preparation, but also it's imported. The energy from the biomasses is still in experimental phase.

The coal is the best researched power source in Macedonia. Total deposits of coal are 906.247 x 10^3t.

The hydro power of the rivers in Macedonia is still not used enough and currently it's exploited in a thermal equivalent of 411 MW. The needs and the participation of every single power source in the energy balance of Macedonia is presented in Table 4.

Type of power sources	Annual consumption (TJ)	Part in the whole consumption (%)
Coal	59.700	56,13
Woods	10.097	9,49
Liquids	33.044	31,07
Renewable	500	0,47
Hydropower	2.498	2,35
Geothermal	510	0,48
Total	106.347	100

Table 4. The needs and the participation of every single power source in the energy balance of Macedonia

7. Hydrochemical analysis of tested water

In the recent period, from geochemical point of view, the best studied are thermo - mineral waters from the springs Ldzi and Kezovica. Further will be presented the hydrochemical characteristics of mentioned system.

Analyses of the chemical composition of the thermo - mineral waters, done on the samples from the exploited well in the spa and the springs Ldzi, showed that in both cases waters are of chloride - sodium, for the springs Ldzi of first type, and for the spa Kezovica of second type.

Total mineralization of the water from the spa Kezovica is 1450.0 mg/l, and from the springs Ldzi varies within 1310.4 to 1507.9 mg/l in the main spring, where in 1953 was drilled the borehole C-4. pH value of this spring and the spa Kezovica is 7.2, based on the examinations of "Industroproject" Zagreb this value is 7.6 for Kezovica and 7.8 for Ldzi Table 5. It is obvious that these waters are on the border between the neutral and alkaline.

About the hardness of the waters from these localities, the general hardness for the gathering on Kezovica is 8.5 dH°, and the carbonate is 9.0 dH°. For the springs Ldzi, the general hardness is 5.9 dH°, and the carbonate is within 7.7 to 9.6 dH°. The largest hardness in the springs Ldzi is in the spring No. 7 where the total mineralization is the smallest. Also, this spring has the lowest temperature (30° C). Probably, it is because of mixing of the thermo - mineral waters with the water from the river Bregalnica, which could be seen on the terrain.

Analyses made from "Industroproject" Zagreb showed increased general hardness - For Kezovica 10 dH°, and for the springs Ldzi 14 dH°.

It should be said that there is no magnesium in the springs Ldzi, and in the gathered waters of the spa Kezovica, the amount of magnesium is 33.0 mg/l and according "Industroproject" Zagreb is 35 mg/l. the question arises why this is present when both waters appeared in the

CATIONS mg/l

Ca++	Mg++	Na+	K+	Fe++	SUM
10.0	35.0	448.0	12.5	1.0	506.5

ANIONS mg/l

SO4-	Cl-	HCO3-	NO3-	NO2-	CO3-	HBO3-	H2AsO4-	SUM
145.0	492.0	166.0	-	-	14.0	120.0	1.0	938.0

ELEMENTS IN

Li	Cs	Br	As	Hg	SUM
0.20	0.12	-	-	-	4.04

SOLVED GASES mg/l

CO2	H2S	NH4	SO2	Cl2	O2	SUM
-	-	0.04	-	-	4.0	—

COLOIDAL SOLVED OXIDES mg/l

Fe2O3	MnO2	SiO2	Al2O3	SUM
-	0.20	3.0	0.04	3.24

TRACES mg/l

Rb	F	J	Sb	Pb
0.15	2.5	-	0.002	0.001

Total mineral. (mg/l)	Dry residue (mg/l)	Lost of ignition (mg/l)	Total hardness (dH)	pH	Eh	t (°C)	Radio-activity (Mah)
1445.0	1400.0	1350.0	10.0	7.6	+210	57	11.57

Curl's formula

$$M1.4 \frac{Cl_{58}HbO^3{}_{17}SO^4{}_{13}HCO^3{}_{11}}{Na_{84}Mg_{12}} t57^0$$

Table 5. Chemical analysis of thermo – mineral waters from spa Kezovica ("Industroproject" – Zagreb)

granites on the fault lines. It should be bearing in mind that thermo- mineral waters Kezovica are on the fault line between granites and Upper Eocene sediments and the springs Ldzi occurred in granites. Magnesium is always present in the sediments which is concluded and from the chemical analyses of the springs with cold water Ribnik (79.2 mg/l magnesium) and Krstot (28.8 mg/l magnesium) Table 6. Also, chemical analyses from Novo Selo showed presence of this element in range of 15.1 - 40.3 mg/l, which means that underground waters from the sediments contain magnesium unlike the waters from the granites.

7.1 Radioactivity

Based on the radioactivity, thermo - mineral waters in the vicinity of Shtip are different, although they are placed in the same rock masses and on the fault lines which through the transverse faults and the tectonic cracks communicate with each other. Based on the analyses of S. Miholik, radioactivity of the waters from the gathering of the waters in Kezovica is 42.82 Mache units (ME), and from Ldzi is 11.57 ME. Based on this, the spa Kezovica is one of the most radioactive spas in Europe.

The increased radioactivity of the water in Kezovica related to that from the spring Ldzi is because of the tectonic, i. e. seismo - tectonic activity of the main fault passing by the Kezovica. This fault is principal seismic dislocation that extends from Strumica, through Kezovica and Sveti Nikole to Kumanovo.

From this arises that with analysis of the radioactivity of dipper underground waters that followed the fault lines could help in exploration of contemporary seismic activity of this terrain.

8. Geological exploration and exploitation of mineral raw materials

Thermal mineral waters by their nature are classified in the group of energy mineral raw materials and are goods of common interest for Macedonia.

According to the Law on Mineral Raw Materials (OGORM no.18/99 and 29/02), a DGE aimed at finding, identifying deposits of thermal mineral water, as well as evaluating their economic effects is conducted by means of a concession.

The Law on Mineral Raw Materials regulates the conditions and the manner of conducting geological explorations, as well as the conditions for exploitation of mineral raw materials situated in the ground or on its surface, concessions for geological research and concessions for exploitation of mineral raw materials, the realization, the maintenance and the utilization of geothermal waters, work safety measures, environmental protection during the exploration and the exploitation of the given area and the geological measurements and plans.

The concession for conducting DGE is issued by the Government of the Republic of Macedonia. The Ministry of Economy, as the competent body responsible for coordination of the procedures for conducting DGE, undertakes inter-sectoral consultations with the state administration bodies with regards to the application for the concession, and primarily with the body responsible for environmental protection and water protection.

Mutual rights and obligations of the Government of the Republic of Macedonia and the concessionaire are regulated with a Concession Contract.

No.	Cadastral num.	Type and location of the object	Water termp. (°C)	pH	Ca++	Mg+	Na++K+	Fe+++	Σκ (mg/l)	SO4--	Cl-	HCO3-	CO3-	NO3-	NO2-	SiO2	Al2O3	ΣA (mg/l)	Total mineralization (mg/l)	Dry residue on 110°C (mg/l)	General hardness (°dH)	Carbonate hardness (°dH)	Class, Group and type of water according Alekin
							CATIONS (mg/l)																
1	1	Drilled well spa "Kezovica"	58	7.2	15.1	33.0	483.3	0.15	467.5	162.4	511.2	195.2	15.5	-	-	75.0	1.2	886.3	1450.0	1419.2	8.5	9.0	Chloride, sodium, second type
2	2	Tharmal well Novo Selo - Ldzi	55	7.2	18.2	-	483.3	0.02	571.5	164.4	538.0	169.0	12.0	Trace	-	62.0	1.0	883.4	1507.9	1480.8	2.5	8.4	Chloride, sodium, first type
3	3	- II -	48	7.0	18.2	-	475.3	0.06	563.5	130.3	545.8	195.2	3.0					874.3	1437.8		2.6	9.0	Chloride, sodium, first type
4	4	- II -	41	7.0	15.1	-	481.7	0.02	496.8	164.8	539.6	164.8	6.0					875.2	1372.0		2.2	8.1	Chloride, sodium, first type
5	5	- II -	47	7.0	18.1	-	467.8	0.06	532.9	158.2	525.4	172.0	4.5					860.1	1400.0		2.6	8.1	Chloride, sodium, first type
6	6	- II -	43	7.0	18.1	-	482.9	0.26	501.2	179.5	543.2	156.7	6.2					885.6	1386.8		2.6	7.7	Chloride, sodium, first type
7	7	- II -	30	7.0	28.4	3.6	455.9	0.06	487.9	165.5	546.4	206.4	4.2					822.5	1310.4		5.9	9.6	Chloride, sodium, first type
8	12	Digged well Novo Selo	13	7.2	102.8	40.3	46.7	0.02	189.8	70.5	67.4	414.8	-					552.7	742.5		23.8	19.0	Hydro-carbonate, calcium, third type
9	9	- II -	15	7.2	21.1	15.1	56.5	0.02	92.7	42.6	33.7	335.5	-					411.8	504.5		15.1	15.4	Hydro-carbonate, calcium, first type
10	10	- II -	14	7.2	84.8	32.4	77.7	0.02	194.9	73.8	61.1	384.3	-					518.2	713.1		19.6	17.6	Hydro-carbonate, calcium, third type
11	13	Spring Ribnik	14	7.0	64.0	79.2	59.1	0.02	202.3	106.6	19.9	579.5	-					705.0	907.3		27.5	26.6	Hydro-carbonate, magnesium, second type
12	15	Spring Krstot	18	7.4	32.4	28.8	105.7	0.13	167.0	47.5	21.3	427.0	-					495.8	662.8		11.2	19.6	Hydro-carbonate, calcium, first type

Table 6. Chemical analysis of the underground waters – spa "Kezovica" and the wider vicinity

Concession for DGE is issued by the Government of the Republic of Macedonia based on a public tender and based on the application/offer submitted by an interested legal entity or natural person. In instances when there are no interested subjects for the public tender, the concession may also be granted based on an offer, which is submitted to the Ministry of Economy.

The applications for issuing a concession is accompanied by a topographic map with a scale of 1:25.000 or 1:50.000 with drawn boundaries of the area where DGE will be conducted or where the exploitation of mineral raw materials will be carried out, as well as a proof of the right to use the results of the DGE, if there have been other prior DGE conducted in the same area

The application for issuing a concession for DGE includes in particular:

- data of the applicant
- type of mineral raw material
- location where that particular mineral raw material is found, in this case geothermal water
- technical and technological explanation for the need to conduct DGE
- Certificate from the Ministry of Finance - Public Revenue Office stating that the applicant has paid all its taxes
- Certificate from the competent Court stating that applicant is not subject to a bankruptcy procedure.

If the area for which a concession for DGE or exploitation of mineral raw materials is being issued, is partially or completely state owned, the Government of the Republic of Macedonia at the request of the concessionaire will grant the use of that land.

A concession for DGE and for exploitation of mineral raw materials can be issued to a domestic or foreign legal entity or natural person (concessionaire) under the conditions stipulated by this Law and the Law on Concessions.

Before the approval of the concession for DGE, the Ministry of Economy will request an opinion from the competent body for environmental protection and protection of nature, the competent body for water protection, as well as other competent bodies.

The legal entity or natural person which has acquired the concession for conducting DGE automatically gains the right to receive a concession for exploitation of mineral raw materials, if the conditions for exploitation of mineral raw materials provided by this Law are fulfilled.

This right is acquired by submitting an application for receiving a concession for exploitation of mineral raw materials at least 6 months prior to the expiration of the concession for DGE.

The concession for DGE is granted for a particular location and for maximum period of 8 years, without a possibility for extension. The location for DGE or exploitation of underground waters or thermal mineral waters can not exceed 2,0 km^2.

The results obtained from the DGE under the terms of a concession are property of the concessionaire and can be sold for an adequate compensation.

Concessions for exploitation of mineral raw materials can be granted for a period of maximum 30 years with a possibility for extension of another 30 years, by signing a new Concession Contract. The application for extension of the concession is submitted at least 3 months prior the expiration of the period for which the concession is granted.

The concession for DGE and exploitation of mineral raw materials can be transferred to another party with the consent from the Government of the Republic of Macedonia. The amount for the concession is determined with the Concession Contract, which is adopted by the Government of the Republic of Macedonia based on a proposal from the Ministry of Economy.

9. Conditions for conducting DGE

Detailed geological explorations (DGE) can be conducted by a domestic or foreign legal entity or natural person, registered for performing the given activity, which fulfils the conditions prescribed by the Law on Mineral Raw Materials and other related laws. During the DGE, all measures for occupational safety and environmental protection must be provided and undertaken. Eventual damages made during the DGE are compensated according to indemnity regulations. DGE and exploitation of mineral raw materials can be conducted on a land or facilities which are public property, and in other areas which are protected by the Law only with a prior approval from a competent body.

DGE is initiated on a defined location based on a previously granted concession and approval for DGE.

The location for the DGE is a part of a terrain circumfered by dots, connected with straight lines and is spreading out to unlimited depth underground between the virtual planes which pass through those lines or natural boundaries which are recorded on the topographic maps, in an appropriate scale.

DGE of mineral raw materials are explorations which provide: detailed information for the hydrogeology of the terrain, the location, the capacity and genesis of the deposit of raw materials, the available deposits of raw materials and the exploitation possibilities accompanied by an assessment of the economic benefit.

The concessionaire has to obtain an Approval for conducting DGE from a competent body in order to start with the on-site works.

10. Approval for conducting detailed geological explorations (DGE)

Approval for conducting DGE is issued on the base of the application submitted by the legal entity or natural person. With the application for issuing of an Approval for DGE the following documents are submitted: Concession contract, Program for DGE, topographic map with the actual border lines of the research area in adequate scale.

The Approval for DGE contains:

- Information about the legal entity or natural person to whom the approval is issued.
- Mineral raw materials which are subject of the DGE
- The area of the DGE with specific coordinates
- The type and scope of the activities that should be performed
- The deadline to start with the exploratory work and
- The duration of the DGE and the deadline for the submission of a performance report on the DGE.
- The holder of the Approval can not start with the DGE, unless he has resolved all property and legal issues for the area defined in the Approval for DGE.
- Approval for DGE is issued within 60 days from the day when the application was submitted.

Against the decision with which the application for the approval is rejected, a complaint can be submitted within 15 days from the day of its receival, to the Committee for resolving administrative procedures of the second instance in the area of economy of the Government of the RM.

The Ministry of Economy can revoke the Approval from Article 33 of this Law, if the legal entity or natural person which is conducting the DGE doesn't start within the determined deadline or if the explorations have been stopped for more than 3 years, unless the reasons for this stoppage were of technical or economic nature, or if the holder of the Approval is not responsible for the reasons for stopping (vis major).

After the stopping or completion of the DGE it's obligatory to file a report to the Ministry of Economy, i.e. an Elaborate for the conducted explorations is being prepared, with complete documentation of the results. The results obtained with the DGE are property of the concessionaire and can be alienated for an appropriate compensation.

11. Exploration of geothermal waters

According to the Law on Mineral Raw Materials which also includes geothermal waters, exploitation of mineral raw materials is an act of obtaining, or releasing mineral raw materials from their natural state (in situ), including the preparatory, accompanying and consequent activities related to the exploitation of mineral row materials. Exploitation of geothermal waters is carried out in a specifically defined space, determined in accordance with the DGE report. The area for the DGE or exploitation of underground waters or geothermal water is maximum 2 km^2.

The right of exploitation of mineral raw materials, which include geothermal waters, is obtained through the granting of a concession for exploitation of geothermal waters.

Before the concession for exploiting is issued, the Ministry of Economy, as a the body responsible for coordination of procedures for exploitation of mineral raw materials, conducts inter-sectoral consultations with other state administration bodies regarding the application for the issuance of a concession, and primarily with the competent body for protection of the environment, waters, forests and cultural heritage, as well as the competent body for transport infrastructure.

The concession for conducting exploitation is granted by the Government of the Republic of Macedonia. The period for which the concession is granted is limited by law to 30 years. The mutual rights and obligations of the Government of the Republic of Macedonia and the concessionaire are regulated with the concession contract.

The issuance of concessions for exploitation of geothermal waters, as well as for other mineral raw materials in accordance with the Law on Mineral Raw Materials (OGoRM no. 19/99 and 29/02) is performed by the Government of R.M based on a public competition or on application/offer of an interested legal entity or natural person.

The application for granting a concession for exploitation of geothermal waters as well as for other mineral raw materials should contain:

- data of the applicant
- proof of the right to use the results of the conducted DGE
- type of mineral raw materials
- location where the mineral raw material is identified, or in this case the geothermal water
- topographic map on a scale of 1:25.000 or 1:50.000 with coordinates of the border lines of the location
- technical and technological explanation of the exploitation of geothermal waters
- Certificate from the Ministry of Finance - Public Revenue Office of paid public taxes by the applicant
- Certificate from the competent Court certificate stating that the applicant is not subject to a bankruptcy procedure and
- Certificate form the competent Court that no bans for performing activities has been issued against the applicant

The exploitation of the geothermal waters can start after the approval for exploration is given to the applicant.

This approval is issued by the Ministry of Economy. For the purpose of granting of the approval for exploitation, the applicant should submit the following:

- Concession contract for exploitation of mineral raw materials.
- Main project for exploitation of mineral raw materials and the deposit, with an expert assessment (review).
- *Elaborate on the environmental impact assessment.*
- Decision for utilization of constructed facilities foreseen in the main project, successively after their construction.
- Proof of resolved property legal relations (this refers to resolved property legal relations with regards to the micro-location of geothermal exploitation)
- Situational plan with defined borderlines of the area where exploitation of the mineral raw materials will be conducted, to the extent it allows defining the boundaries of the area, as well as defining public and other facilities.
- Transfer of concession for carrying out the exploitation of mineral raw materials (geothermal waters) can be carried out in accordance with the Government of the Republic of Macedonia.

The Ministry of Economy can revoke the Approval for exploitation from Article 63 of the Law on Mineral Raw Materials (according to the Law on Changes and Amendments to the Law on Mineral Raw Materials, OGoRM no. 29/02) if the performer of exploitation of mineral raw materials doesn't start within 3 years of the day of the issuance of the Approval, or if regular exploitation was stopped for more than 3 years, unless the reasons for the stoppage were of technical or economic nature, or if the holder of the Approval was not responsible for the reasons for the stoppage (vis major).

Supervision over implementation of the Concession Contract is carried out by the Concedent, i.e. the Ministry of Economy and the State Inspectorate for Technical Inspection through mining and geological inspectors.

The amount of the concession fees is determined with a special Decision for determining the criteria for the amount of the concession fee for conducting DGE and exploitation of mineral raw materials, adopted by the Government of the Republic of Macedonia.

It's important to note that the calculation of the concession fee is done per production unit of produced mineral raw material.

For a complete overview of the administrative procedure for concessions for geological explorations and exploitation of mineral raw materials, including geothermal water, a brief presentation of the required documents and the procedure is provided below.

12. Commentary instead of conclusion about the legislation for issuing concessions for exploiting the geothermal energy in the Republic of Macedonia

Geothermal resources (geothermal waters and their geothermal energy) by their nature are classified in the group of energy mineral raw materials and are goods of common interest of the country, according the Law on Mineral Raw Materials (OGoRM no. 18/99 and 29/02). Detailed geological explorations (DGE) aimed at finding; identifying deposits with thermal mineral water, as well as evaluating their economic effects are conducted through concession. Concession as an institute in legislation entails obtaining a certain right, with or without certain conditions, for a limited or unlimited time. A Concession for DGE, as well as Concession for exploitation of mineral raw materials, can be issued to a domestic and foreign legal entity and natural person (concessionaire) under the conditions defined by the Law on Mineral Raw Materials and Law on Concession. With regards to geothermal energy, the experiences in this area show that two types of concession are characteristic: the right to perform geological explorations in order to identify the energy resource and right to use the resource under specific conditions. Unfortunately, this is where all similarities with mineral raw materials end, because the concessions mostly apply to using hot water or steam, which automatically includes the application of the Law on Waters. Because of the potential negative environmental impact caused by the use of hot water and water with high mineral content, the legal provisions related to environmental protection are applied in conjunction with the Law on Ecology.

This immediately poses the question whether is it possible to harmonize three different legal areas in order to ensure more efficient resolution of the problem, and if this is not possible, is it necessary to create a special legislation for utilization of geothermal energy, which will define all aspects involved in geological exploration and exploitation of this specific energy resources. Concessions are granted by the Government of the Republic of Macedonia, based on a submitted application.

There is a clear difference in the legislations of European countries regulating the obtaining of licenses for exploration and exploitation of the geothermal resources, as a consequence of the specific development of the legislations in general and the specificities of the geothermal resource themselves. However, the application in practice of certain provisions is common for all European countries, and the Republic of Macedonia should strive and move in this direction, as described below:

Ownership of the energy resource

Geothermal energy resources are owned by the state and can not be alienated. The ownership of the energy resource is in no form related to the ownership of the land under which the resource is located. The former means that the owner of the land has no right to obstruct any investigations or exploitation of the energy resource, regardless of the use of the land, but in the same time means fair compensation for possible damages made in the course of exploration, building the infrastructure or exploitation. The owner of the land is given priority in applying for concession for DGE or exploitation if he offers the same or approximately the same conditions as the most favourable offer, because the concessions are issued through a public tender.

Conducting of exploratory work for identification of the energy resource

The state issues Approval for DGE under strictly defined procedures and conditions only to qualified companies and institutions which have to provide technical documentation and bank guarantees, ensuring its implementation. The role of the state is to control and provide full protection of the energy resource and the environment where the exploration is carried out or to prevent speculations aimed at stopping the activities for activating the energy resource.

Exploitation of the geothermal resource

The state grants the rights for exploitation of the geothermal resource under strictly defined procedures and conditions. The license for exploitation of the energy resource entails full protection of the energy resource from excessive or improper usage. The right to control also means accepting direct responsibility to the holder of the concession i.e. the state can not issue concession for geothermal fields for which it has already issued concessions to specific users.

In the procedure for issuing concessions for exploration and exploitation, it is important to emphasize the system of dual responsibilities i.e. full protection of the rights of the state, but also the rights of the holders of the license for exploration and exploitation of the energy resource. The more they are defined by law in a clear and directly applicable manner, the more their practical application will be efficient.

All these aspects in the development of the legislation regulating the issuance of concessions for exploration and exploitation, and consequently the utilization of the geothermal energy, are mostly applied in various European countries that have a tradition in the utilization of this specific resource and can provide useful insights for the development of a concrete legislation in the Republic of Macedonia.

The implementation of the current Law on Mineral Raw Materials in practice, as well as the taking of responsibilities as a result of the need for approximation of Macedonian with European legislations, lead to the need to draft a new Law on Mineral Raw Materials. The law proposal was drafted by the Ministry of Economy in collaboration with the German Technical Cooperation (GTZ).

The main characteristic of the new law proposal is that a much clearer systematization of the provisions is applied as compared to the existing one. Also, the proposal for adoption of the Law incorporates Directives which are indirectly transposed in this Law through the Law on Environment.

Other specific new feature of the text of the law proposal is the introduction of the National program for development and planning in the field of mineral raw materials. This National program defines the scope, goals and guidelines for the geological exploration, as well as the sustainable utilization of mineral raw materials, while respecting the specificities of different areas, the features and distribution of mineral raw materials, and the need for exploitation of mineral raw materials which are crucial for the development of the economy of the Republic of Macedonia.

The new text of the law also proposes that the rights for DGE is to be obtained through an approval that will be issued by the competent body for management of mineral raw materials (until now the right to conduct DGE was obtained through the granting of a concession for DGE). Another novelty in the law proposal is the possibility for transfer of the license for DGE another person, but under the same conditions. Also, the rights from the finished DGE can be transferred to a third person, which is in accordance with the principles of market economy. The right to exploit mineral raw materials will be obtained through the granting of a concession for exploitation of mineral raw materials. There are two ways for the granting of a concession for exploitation of mineral materials which are foreseen. The first one is through a public tender in cases which are stipulated in the National Program for Development and planning in the area of mineral raw materials, as well as when the Republic of Macedonia possess revised elaborates of conducted individual DGE. In accordance with market economy principles, the text provides full or partial transfer of the concession for exploitation of mineral raw materials under strictly defined conditions. The criteria, amount and the method of payment of the concession fees will be determined by the Government of the Republic of Macedonia, in accordance with the principles of market and open economy.

Another important new feature is the allocation of funds received from concession fees.. This law proposal stipulates that half of the funds from concession fees for exploitation of mineral raw materials (which includes thermal mineral waters) are to be transferred to the Budget of the Republic of Macedonia, and the other half to the budget of the Municipality where the exploitation takes place (50%-50%). It also introduces the obligation for rehabilitation of damages caused by the operations of the entity in the course of exploitation of mineral raw materials. The main intention of the new law is to emphasis the commitment to protecting the environment. Consequently, it imposes an obligation on the performers of the DGE and exploitation of mineral raw materials to comply with the requirements of this law and the Law on Environment and it reinforces the penal policy in the area of exploitation of mineral raw materials and geothermal waters as a mineral raw material.

13. Conclusions

1. Geothermal energy is not a "new" and unknown source of energy in our country. The same is proven in the past with its economic benefit and it is competitive to any other kind of energy source in Macedonia.
2. In eastern and eastern Macedonia, at the moment are registered over 30 springs and occurrences of mineral and thermo-mineral waters with maximum capacity of more than 1.400 l/s and evidenced deposits for exploitation of around 1.000 l/s with water temperatures from 20-79°C, with significant quantities of geothermal energy.
3. The calculation of the valorization of the total balance of the geothermal exploitation reserves in eastern and south-eastern Macedonia (standing 2006) shows: maximum available power is equal to 173 MW or capacity for annual production of 1.515.480 MWh/ year as heat equivalent.
4. Eastern Macedonia has numerous occurrences of thermal and thermo-mineral water that are used in spas and recreational medical centers in the balneology sphere. The geothermal energy from this thermo-mineral water is used in four existing thermal spas with total yield of 250 l/s, total caloric power (heating power) of 35 MW and annual exploitation of 112.5TJ.
5. Taking in consideration that the existing production capacities are based either upon natural springs or shallow boreholes, the maximal current capacity can not give the actual picture for the real capacity of this energetic resource. Based on the existing data it is evident that if supported with relatively small investments in the already started exploration works of these potentials during the 80's of the past century, the capacity could be easily doubled. By providing bigger support it could help to increase and multiply the current capacities over 4-5 times in the next few years. Anyway, the amount of the investments needed for exploration and exploitation of geothermal water wells, points out that without direct support from the state it is not possible to enjoy the benefits of this energy. That is why the Government should treat the geothermal resources as energy of special importance.
6. Further explorations of the geothermal energy above all requires changes in the attitude toward this kind of energy and the dealing with the legislation in this area, which will help to reach a higher level in the utilization of the geothermal waters. This is especially applicable for the hydro geothermal waters with temperatures lower than 40°C that still are not used as energy source. A question which could be imposed here: Is it necessary to adopt special legislation for utilization of the geothermal energy which will define all the aspects in the field of geological explorations and the utilization of this specific energetic resource.
7. Future exploration works in eastern and south-eastern Macedonia should be focused in the already registered areas where there is an indications for increased terrestrial

thermal flow (yield), and registered superficial manifestations from the previous geothermal explorations like in the Delchevo region, broadening of Vinica and Kochani geothermal field, Probishtip region, Radovish region, Dojran - Valandovo-Gevgelija region, in the Vardar zone and the contact-part with the Serbian-Macedonian massif in Bregalnica-Strumica region.

14. References

Alekin, O. A., 1953: Basics of Hydrochemistry, Leningrad

Arsovski, M., 1997: The tectonics of Macedonia. Faculty of Mining and Geology - Stip. Page 300.

Arsovski, M., Stojanov, R., (1995): Geothermal phenomenon related to the Neotectonics and the magmatism in the area of R. Macedonia. Special edition, MANU, Skopje.

Bonchev, E., 1974: Kraishtides. Tectonics of the Carpathian ' Balkan Regions, Bratislava.

Dimitrievic, M., (1974): The Serbo-Macedonian Massif. The tectonics of the Carpathian-Balkan Regions. p.p. 291-296, Bratislava.

Georgieva, M., 1995: Geothermal resources in the area of Vardar and the Serbian – Macedonian mass in the territory of Macedonia (Doctoral dissertation) Faculty of Mining and Geology – Stip page 190.

Kekic, A., Mitev, Z., 1976: A report on the thermo-mineral waters of the bath Kezovica and Ldzi with a special review on the drilling process in the bath in 1975-1976. Professional fund of the Institution of Geology – Skopje.

Kotevski, G., 1977: Hydrogeological research on minerals and thermal waters on the territory of S.R. of Macedonia (Final analysis). Professional fund of the Institution of Geology – Skopje.

Kovacevic, M., Petkovski, P., Temkova, V., 1973: Explanation for sheet Delcevo, BGM SFRY, M 1:100 000, Geological Institute, Skopje.

Miholic S. 1953 – Previous report on the research on the thermo-mineral waters of the Stip spa and the acid waters in Bogoslovec. Professional archives of spa Kezovica - Stip

Miholic S. 1953 – A report on the research on the thermo-mineral waters in the spa Stip (spring Ldzi). Professional archives of spa Kezovica - Stip

Naunov, J., 2003: Geotermal system Geoterma. Second conference for geothermal energy in Macedonia (Proceedings). Bansko. Pages 5-15.

Popovski K., Micevski E., Popovska-Vasilevska S., 2005 Macedonia – Country Update 2004, Proceedings, World Geothermal Congress, Antalya, Turkey, 24-27.

Rakicevic, T., Kovacevic, M., 1973: Basic geological map of SFRJ, sheet Delcevo, M 1:100 000, Geological Institute, Skopje.

Serafimovski, T. 2001: The relationship between geothermal systems and hydrothermal mineralization systems. First Counceling on geothermal energy in Macedonia., Bansko, Proceedings.

Vidanovic M. 1955 – Proposal for central therapeutic facility in the bath Kezovica. Professional archives of spa Kezovica – Stip.

Vidanovic M. 1955 - Anticipating indications for treatment and the use of thermo-mineral waters in the bath Kezovica. (A foundation for building bath facilities). Professional archives of spa Kezovica – Stip.

4

Environmental Impact and Drainage Geochemistry of the Abandoned Keban Ag, Pb, Zn Deposit, Working Maden Cu Deposit and Alpine Type Cr Deposit in the Eastern Anatolia, Turkey

Leyla Kalender

Firat University, Department of Geologycal Engineering, Elazig
Turkey

1. Introduction

This study includes the effect on the environment of abandoned and working mine deposits which are lead, zinc, silver and copper deposit in Keban, working copper deposits in Maden and chromite deposit in Alacakaya. These are the largest metallogenic province of Turkey (Figure 1).

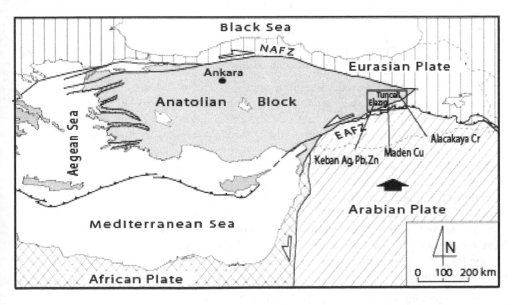

Fig. 1. Location map of the study area.

The Keban region was one of the most important base metal areas of eastern Turkey. The Keban area has been an important mining area since B.C. Archaeological studies revealed that ancient miners have mined the Keban area for gold, copper, lead, silver and iron. During twenttieth century, the mining activity was restricted to iron, lead, zinc, silver and fluorite. The Keban mine was operated for Pb and Zn production at various times between the 1940s and the late 1980s. There were several prospect pits and mining tunnels in the area. However, mining ceased in 1988 and the site was abandoned. The west and east Euphrates slag sites account for on average 2.16 ppm Au, 66.41 ppm Ag, 2.38 % Pb, and 0.52 % Zn (Kalender & Hanelçi, 2001). The Euphrates River and its tributaries, in particular, Karamağara Stream, drain the former metal-mining areas, which have been contaminated by the release of heavy metals from old mining operations within the Keban mining district. Many geological studies have been conducted around the Euphrates River (Kalender & Hanelçi, 2001; Kumbasar, 1964; Kineş, 1969; Zisserman, 1969; Köksoy, 1975; Kipman,1976; Akıncı et al., 1977; Balçık, 1979; Yılmaz et al. 1992; Çelebi & Hanelçi, 1998; Çalık, 1998; Kalender, 2000; Kalender & Hanelçi, 2001; 2002, Bölücek, 2002). However, none of these studies have addressed the issue of environmental contamination in the area. In this study will be presented geochemical results from several sampling media (sediments, mine-drainage waters, spring waters, river waters, moss, and algae) from the Keban mining district.

Maden copper deposit was important ore deposit of Turkey. Mining activity had been operated from 4000 B.C. up to the recent (Seeliger ve diğ., 1985, Tızlak, 1991). The mine was run by Rome, Seljuk and Ottoman Empires (1860-1915) and the Republic of Turkey. Maden copper mining operations have been transferred with the establishment of Etibank in 1935. Due to exhaustion of economic reserves were decided to close the facilities in 1995. However, Ber-Oner Mining has operated of the waste from 1995 to 2005. Since 2007, Maden copper mining waste has been operated by Eti Holding and SS Yıldızlar Holding and produced 20 000 tons of copper a year. Many geological studies have been conducted around the Maden copper deposit area (Erdoğan, 1977; Özkaya, 1978; Özdemir & Sağıroğlu, 1998; Özdemir & Sağıroğlu, 2000; Kırat et al, 2008).

In this study will be stressed dissolution methods and metal dispersion patterns were determined suitable for such an area by stream sediment survey. The physical and chemical feature of the area indicate that the metalic contents of the stream sediments originate from physical events rather than chemical events. In this case, 180-106 μm sediment fraction were analysed. In additon to surface and seepage water samples were collected from the area and several elements were analyzed and data were evaluated in term of pollution. After study was conducted to evaluate the influence of anion (sulfate) and metals (copper, iron, manganese, zinc and nickel) on the reduction by these bacteria. Furthermore, the sulphate reducing bacterium methods were evaluated to selectively precipitate metals and reduction sulphate from Maden (Elazığ) copper Deposits AMD seepage waters in two samples locations. Initial and final metals and sulphate concentration before and after experiments, were measured and finally evaluated influence of bacteria.

Guleman ophiolite made up of tectonites which comprises dunite and chromite bearing hazburgites and cumulates which contain dunites, wehrlite, clinopyroxenite, gabbros, diabase dykes, sheetdyke complex and basic volcanites are other constituents of the ophiolite. This area has been studied by Erdoğan, 1977; Erdoğan, 1982; Başpınar, 2006. There are thirty rock samples of Guleman Ophiolite were analyzed by Başpınar in 2006 and ten groundwater samples were analyzed by Kalender in 2010. This study includes correlation of both analysis results of data.

This study mainly focus on the concentrations and distributions of heavy metals and potentially toxic elements in the various sampling media impacts on environment pollution in the three different ore deposit

2. Keban Ag-Pb-Zn deposit

2.1 Sampling and analytical methods

Water, stream-sediment, algae samples were collected from the study area. Water samples were taken from natural springs (LK 1, LK 2, LK 3, LK 4, LK 6, LK 7, LK 8, LK 9, LK 10), mine-drainage waters (LK 5: fluorite production gallery; LK 8: Pb-Zn underground-mine production workings) and the Euphrates River (LK 11). Algae samples were collected from the bed of the Euphrates River (AK 13, AK 14, AK 18) and only one moss sample. Sediment samples were collected from the Euphrates River bed (LZ 120, LZ 150, LZ 160, LZ 170), from Karamağara Stream (LZ 270, LZ 310, LZ 330, LZ 340, and LZ 350) and from precipitates at the gallery mouth (LZ 370) (Figure 2). All of the sampling work was done in June 2002 (Kalender & Bölücek, 2004; Bölücek, 2007).

Water samples were collected into 250-ml polyethylene containers and filtered through 0.5 µm membrane fitler paper (4.5 cm diameter). Temperature and pH measurements were conducted at the sampling sites. 5 ml concentrated HNO_3 (MercTM ultra pure) was added to the samples for metal analyses. Samples were stored in the laboratory at +4°C until they were analyzed. For each sediment sample, 2 kg of sediment was sampled from a depth of about 10 cm by sieving to 2 mm mesh size. For analyses, samples dried in the laboratory were sifted using stainless steel sieves to mesh sizes between –80 and +140. The samples were subjected to partial-digestion involving the use of a cold solution of 0.3 % NaCN and 0.1 NaOH to extract weakly-bound elements from clay, organic matter, and amorphous Mn and Fe hydroxides (Kelly et al., 2003). Green algae *Cladophora glomerata* in sample AK 14 is a typical species in rivers; it is found together with macroalgae (Entwisle,1989; Power, 1992). *Cladophora glomerata* is found mainly in volcanic areas (Whitton et al.1998). The presence of *Cladophora glomerata* is used as an indicator of good-quality water. Aksın et al. (1999) examined the Keban stream algae and determined 70 taxa from the *Clorophyta, Cyanophyta, Bacillariophita,* and *Dinophyta*. These workers reported that diatoms are the dominant group of organisms among the pelagic and benthic algae population because of the rapid water flow of Keban Stream. Algae species living in the Euphrates River differ from those in Keban Stream due to the temperature and low current velocity in the former. They are quite common in the river beds having low current velocity, that receive little light, and that have temperatures above 10°C (Schönborn, 1996). The temperature of the Euphrates (12.4–12.7°C) is suitable for the growth of this type of algae. Investigations indicate that the amount of this organism will increase with increasing temperature. The size of *Didymodon tophaceus* is around 0.5–5 cm. It is composed of circular cells that widen toward the base, and is mainly

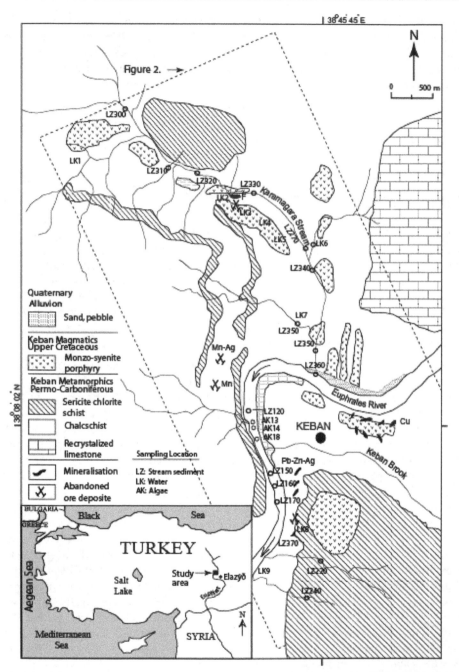

Fig. 2. Simplified geological map of the study area Kalender & Hanelçi, (2001; 2002) and locations of sampling sites (natural spring waters: LK 1-LK 4, LK6, LK 7, LK 9, LK 10; mine-drainage waters: LK 5, LK 8; Euphrates River water: LK 11).

olive green but turns brown in its lower parts. Moss and algae samples were washed using distilled water and dried in air in the laboratory. 0.5 gr sample batches were digested to dryness at a temperature of 95°C for one hour after adding 2 ml of HNO_3 and then 3 ml of a 2:2:2 $HCl:HNO_3:H_2O$ mixture. A total of 38 elements in the water samples were analyzed by ICP/MS (inductively coupled plasma/mass spectrometry); F in water was analyzed by ion electrode. Sediment samples were analyzed by ICP/MS, except for Au and Hg which were determined by atomic absorption (AA) after MIBK extraction and flameless AA, respectively. The algae and moss samples were analyzed by the ICP/ES (inductively coupled plasma emission spectrometry) and ICP/MS methods.

2.2 Results and discussion

It was determined that Mg, Ca, S, Fe, F, Mn, Zn, Mo, Ba, Pb, U, Ni, Cd, Co, Ag, Cu, Sb and Se contents of groundwater are above the standards of drinking water and it is also enriched in some toxic elements such as Al, Cd. Cr, Fe, K, Mg, Mn, Na, Pb, Sb, SiO_2, Tl and Zn. The springs particularly issuing from the mine galleries are probably the main source of pollution parameters. Analyses of weak leaching elements of -80 +140 mesh size sands collected from some main river and creeks indicate high As, Cu, Zn, Mo and Ag contents, various algae samples are characterized with high Cu, Pb, Zn, Ag, Cd and Cr concentrations. In this study will be determined the source and impact on environment of pollution in water, stream sediment and plants.

Element concentrations in different water samples are variable, with the Ba, Ca, Cd, Ce, Co, Cr, Cs, Mg, Mo, Pb, Rb, S, Se, U, Zn contents of most of these samples being above the world averages encountered in ground water (Table 1). The high metal concentrations in the samples originated from waters moving through the old mine adits or unexploited mineralized areas. This conclusion is supported by the compositions of waters collected from the adits (LK 5, LK 8).

In Table 2, the results of chemical analyses are compared to various standards in order to evaluate the drinking quality of waters. The concentrations of Cd, Pb, Tl, Cr, and Sb measured in these samples were higher than 0.005 mgl^-., 0.015 mgl^-., 0.002 mgl^-., 0.05 mgl^-., 0.006 mgl^-., respectively, which exceed the limits allowed for drinking waters (USEPA, 2002; WHO, 1993; Canada MAC, 2001; EEC, 1992). According to international standards, the concentrations of some other elements (Al, Fe, K, Mg, Mn, Na, SiO_2, Zn, and F) are significantly higher than the maximum permissible levels and, therefore, these waters should not be used for drinking. In light of these data, it is concluded that contamination of these waters by heavy metals and other toxic elements through mining, production galleries, and slag sites also causes environmental contamination.

Spearman correlation coefficients were calculated and their values at a significance level of $\alpha=0.05$ are summarized in Table 3. Spearman correlation coefficients given in brackets for the element pairs are: As–Cs (0.68), As–Re (0.65), Cd–Co (0.75), Cd–Cu (0.64), Cd–Fe (0.66), Cd–Mn (0.89), Cd–Zn (0.88), Se–S (0.85), Se–Re (0.86). Se, a chalcophile element found in appreciable amounts in polymetallic sulfide ores, easily dissolves in water by oxidation of sulfides and shows significant correlation with S and Re.

Elements	Median	Mean	Detection Limits	St. Dev.	Maximum	Groundwater World Average
As	2.5	3.3	0.5	2.91	11	2
B	20	27.4	5	13.23	58	10
Ba	43.345	44199	0.05	24.21	76.82	20
Br	46	124	5	195	654	20
Ca	89825	163275	0.05	154884	455325	50000
Cd	0.115	2.475	0.05	5.17	14.82	0.03
Ce	0.115	0.504	0.01	1.2	3.9	-
Cl	4000	10800	1	14413	49000	20000
Co	0.11	2.219	0.02	4.89	14.82	0.1
Cr	12.75	53.96	0.5	127.79	416.3	1
Cs	0.43	0.71	0.01	0.82	2.48	0.02
Cu	2.55	3.64	0.1	3.03	10.2	3
Fe	31	166	10	356.36	1171	100
In	47.5	45.3	0.01	7.85	57	7
K	2745.5	6410	0.05	8189	27565	3000
La	0.065	0.74	0.01	2.09	6.69	0.2
Li	6.5	13	0.1	14.32	41	3
Mg	27742	73133	0.05	94490	307178	7000
Mn	4.64	4421	0.05	13559	43000	15
Mo	15.9	34.38	0.1	39.96	109	1.5
Na	4228.5	15716	0.05	26916	90025	30000
Nd	0.055	0.18	0.01	0.38	1.25	-
P	22	24.7	20	5.4	33	20
Pb	10.75	15.86	0.1	14.58	47	3
Rb	14.54	21.94	0.01	24.89	82.28	1
Re	0.04	0.46	0.01	1.19	3.84	-
S	36000	138700	1	176251	399000	30000
Sb	0.575	1.49	0.05	2.24	7.26	2
Sc	5.135	5.38	1	1.49	8.29	-
Se	1.15	2.05	0.5	2.07	7.5	0.4
Si	19	18.5	40	5.43	29.6	16
Sn	0.425	0.992	0.05	1.26	4.3	0.1
Sr	1082.63	2008	0.01	2125	5668	400
Tl	1.31	4.36	10	8.51	27.75	0.002
U	17.58	17.71	0.02	14.08	32.02	0.5
Y	0.04	0.15	0.01	0.33	1.09	-
Zn	101.55	2466	0.5	5881	18573	20
F*	-	-		-	2200	100

Table 1. Statistical parameters for chemical composition of water samples (n=10, μgl^l-, * mgl^l-). Elemental contents in groundwater are from Rose et al., 1979; Concentrations higher than the normal values for water are shown in bold.

Elements	Median	Mean	Std. dev.	Guidelines	Maximumum values and sample code
Al	0.064	0.64	1836	0.05-0.2*	5.872 (LK-5)
Cd	0.0001	2.475	5.17	0,005*	0.014(LK-5);0.009 (LK-8)
Cr	0.013	0.054	127.79	0.1*	0.416(LK-2)
F	-	2200	-	1.5**	2200(LK-5)
Fe	0.031	0.166	356.36	0.3*	1.171(LK-8)
K	2.745	6.41	8189	12****	27.565 (LK-5)
Mg	27.742	73.133	94490	50****	307.178 (LK-9);163.114(LK-8);88.830 (LK-5)
Mn	0.005	4.421	13559	0.05*	43 (LK-5);0.1(LK-4);1.1(LK-8)
Na	4.228	15.716	26916	20*	90.025(LK-9);243 (LK-8)
Pb	0.011	0.016	14.58	0.0 15*	0.047(LK-10);0.018(LK-5);0.019(LK-2); 0.034(LK-1)
Sb	0.006	0.001	2.24	0.006*	0.007(LK-8)
SiO$_2$	19	18.05	5.43	10***	29.6(LK-5)
Tl	0.001	0.004	8.51	0.0005-0.002*	0.028(LK-5);0.0065(LK-4);0.002(LK-8);0.004(LK-3)
Zn	0.101	2.466	5881	5*	18.573(LK-5);5.2(LK-8)

Table 2. According to different drinking water guidelines that elements exceed of maximum contaminant level (MCL), n=10 mgl^{l-}. *= USEPA: United State Environmental Agency **= WHO: World Health Organization Guidelines, ***= Canada MAC: These limits are established by Health Canada; ****= EEC: Europium Economy Community.

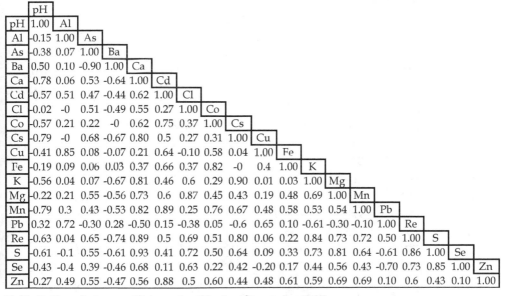

Table 3. Correlation coefficients ($n=10$); significant at $\alpha=0.05$.

The negative correlation (–0.70) between Pb and Se indicates different solubility behavior of the Se and Pb under surficial oxidizing conditions. The negative correlations between Pb–S (–0.61) element pairs may indicate that Pb compounds with S (e.g., lead sulfo-salts) are quite sensitive to dissolution. However, their enrichment in the waters is attributed to formation of complex Pb–Mn and Pb–Cr oxides (oxyhydroxide complexes). Negative correlation between Al and pH (Table 3) indicates that the amount of aluminum dissolved in water increases with increasing acidity. As seen in the underground spring, the presence of CaF_2 in the water affects Al solubility by forming $AlF_6{}^{-3}$ type complexes.

The elemental contents of water, syenite porphyry, fluorite, molybdenite, silver-manganite and stream sediments are given in Table 4. To date, no analyses have been done for As, B,

Elements	**Fluorite molibdenite	**Silver manganite	***Syenite porphyr	Detection Limits	Stream Sediment Max.	Min.	Mean	Detection Limits
Al	47900	2700	46300	100	<10	<10	<10	%0.01
As	168	328	-	1	99999	0.245	13.712	0.1
Ba	10600	774000	2516	5	1.57	0.01	0.67	0.5
Br	11.64	21	-	0.1	<5	<5	<5	5
Ca	250000	148400	26200	100	20	5	12	5
Cd	16.75	3	-	0.1	-	-	-	-
Cl	-	126	-	0.1	85	10	29	10
Co	7.35	6	9.86	0.2	0.061	0.002	0.021	0.1
Cr	33.05	34	-	0.5	0.12	0.05	0.08	0.5
Cu	163	82	173	0.1	13.57	0.3	6.1	0.01
Fe	22800	84800	5200	100	12	5	5.86	0.01
K	4900	950	73400	100	172	19	59	0.01
Mg	18900	76600	5800	100	<5	<5	<5	0.01
Mn	3200	51900	3300	100	0.5	0.1	0.33	1
Mo	2500	-	-	0.1	12.75	13	2.94	0.01
Na	1500	270	15500	100	-	-	-	0.001%
P	200	43600	20	10	3	1	1.44	0.0001%
Pb	5500	6100	98.18	0.1	0.12	0.01	0.03	0.01
Rb	89	15	181	0.1	-	-	-	-
S	2500	3700	-	5	-	-	-	-
Sb	89	61	-	0.1	0.04	0.005	0.014	0.02
Se	25.84	7	-	0.5	1.34	0.07	0.41	0.01
Si	181200	35800	617800	100	-	-	-	-
Sn	45.55	65	-	1	-	-	-	-
Sr	700	-	673	0.5	2.47	0.02	1.4	0.5
Tl	4.08	10	-	0.1	-	-	-	0.02
Zn	872	456	163	1	14303	0.38	1665	0.1
U	49.71	6	-	0.1	-	-	-	0.05
F(n=4)	183000	571	-	100	6120	460	2542	100

Table 4. Content of elements in mineralizations and siyenite porphyr and stream sediments in ppm (**= Çelebi ve Hanelçi,1998, n=21, ppm; *** Kalender, 2000, n=22, ppm).

Br, Cd, Ce, Cl, Cs, La, Li, Mo, Nd, S, Sb, Sc, Se, Sn, Tl, U, and F in the syenite porphyries. The F, Mo, Mn, and Ag in the water samples were derived from fluorite (CaF_2), molybdenite (MoS_2), pyrolusite (MnO_2) and Ag-sulfosalts while their Cd, Co, Cr, Cu, Fe, Sb, Se, Sn, and Z contents originated from dissolution of chalcopyrite ($CuFeS_2$), arsenopyrite (FeAsS), enargite ($Cu_3AsSbS4$), and loellingite ($FeAs_2$). The Na, Al, Ca, K, and SiO_2 concentrations of the waters resulted from the dissolution of the syenite porphyries themselves (Bölücek & Kalender, 2009).

Bioaccumulation values were calculated from two algae and one moss species. Bioaccumulation is a parameter that describes bioconcentration as the ratio of the concentration of a chemical species an organism to the concentration in the surrounding environment. Bioaccumulation factor (BAF) is calculated as follows:

BAF = (whole body concentration of X) / (exposure media concentration of X) The element contents and BAF values of the species are given in Table 5. *Bangia atropurpurea* (red algae) in the samples (AK 13 and AK 18) from the Euphrates river bed (T= 13 Co, pH = 8) was first described by Kishler & Taft (1970), in Lake Erie. This species is generally common in marine environments, gulfs and bays (estuarine). However, it is also found in continental waters where climatic conditions are suitable. *Bangia atropurpurea* is commonly found in ion-rich alkaline (pH= 8–8.7) fresh waters. When compared to *Cladophora glomerata*, it has a higher capacity for absorption of Cu, Pb, Ni, Mn, Au, Cr, Mg and S. *Bangia atropurpurea* contains 6 fold Cu, 26 fold Ni, 2.7 fold Mn, 11.5 fold Au, 20.5 fold V, and 77 fold Cr compared to *Cladophora glomerata*. As is well-known, bioaccumulation is the net accumulation of a chemical by an aquatic organism as a result of uptake from all environmental sources. Thus, BAF values in this study were calculated for green and red algae samples according to the elemental contents of river water and, for moss samples, according to the elemental contents of spring and mine-drainage waters collected from the same locations (Table 5). The chemical concentrations in tissues of aquatic organisms and water can be defined in terms of chemical partitioning between different biological or chemical phases. BAF values in *Bangia atropurpurea* for Mo, Cu, Ag, Sb, P, Mg, Ba, and SO_4 are higher than those *Cladophora glomerata*, wheras BAF values for Zn, Co, V, Ca, La, B, Al, Hg, and Se are higher in *Cladophora glomerata* than those in *Bangia atropurpurea*. Total BAF (total concentration of the chemical in tissue / total concentration of chemical in water) values for *Bangia atropurpurea* are higher than those for *Cladophora glomerata* (Table 5). In sample AK- 14 in comparison to *Bangia atropurpurea*, *Cladophora glomerata* has higher capacity of absorption for Cd, Bi, B, Hg, Se, Fe, Co, Ag, and Zn (Table 5). Particularly, the Hg (13 fold) and Cd (3.8 fold) contents of *Cladophora glomerata* may indicate that these algae may have a direct effect on the ecosystem. The moss, *Didymodon tophaceus (Brid.) Lisa* generally grows on Paleozoic limestones and dolomites, and is generally observed along banks and hilly areas close to river bed with gentle slopes. The highest BAF values in this study were observed in moss (*Didymodon tophaceus*) sample MK 20, indicating that the moss tissues have an ability to absorb heavy and toxic metals from mine-drainage waters. In this study, it was observed that there is considerable pollution of the waters and stream sediments in the Euphrates River. However, the presence of water plants have a positive effect on the aquatic ecosystem by absorbing the heavy metals and toxic elements (Pb, Hg, Cd, Cr, and As). Therefore, algae and moss can be used to remove some of the heavy metals and toxic elements from mining-waste water.

Elements	Detection Limits	LK13 Bangia atropurpurea	LK 14 Cladophora glomerata	LK 18 Bangia purpurea	LK 20 D. Tophaceus (Brid.) Lisa	LK 8 Derebaca Gallery	LK 11 Eupharate River
Mo	0.01	0.38	0.26	0.33	0.36	0.001	0.003
Cu	0.01	13.15	1.54	9.26	68.61	0.002	0.003
Pb	0.01	1.75	<0.01	1.85	269.76	0.004	0.0002
Zn	0.1	155.4	267	117.9	3051.3	5.19	0.003
Ag*	2	11	70	8	1575	-	<0.05
Ni	0.1	6.3	0.1	2.6	25.1	-	<0.0002
Co	0.01	0.49	0.65	0.28	6.96	0.007	0.000008
Mn	1	22	15	41	1034	1.076	<0.00005
Fe	0.001%	730	930	390	3450	1.171	<0.01
As	0.1	6.3	13.9	13.3	123.5	0.011	0.003
U	0.01	0.27	0.39	0.15	6.86	0.04	0.0074
Au*	0.2	3.3	<0.2	2.3	17.3	<0.05	<0.05
Th	0.01	0.1	0.11	0.04	0.03	-	<0.00005
Sr	0.5	41.4	53.3	56.6	307.8	2.181	0.842
Cd	0.01	0.21	0.57	0.15	12.22	0.009	<0.00005
Sb	0.02	0.03	<0.02	0.19	0.98	0.007	0.001
Bi	0.02	0.05	145	0.02	0.09	-	<0.00005
V	2	11	226	11	12	-	0.001
Ca	0.01%	10400	13800	9800	46900	436.67	55.37
Pb	0.01	3020	840	4210	2060	0.02	0.023
La	0.01	0.29	2.21	0.12	1.02	0.0003	0.00001
Cr	0.1	4.98	0.05	3.87	3.87	0.011	<0.0016
Mg	0.0001%	2130	1830	2080	2091	163.11	18.56
Ba	0.1	8.2	22.3	12.6	30.3	0.016	0.026
Ti	1	14	13	14	7	-	<0.01
B	1	12	347	9	336	0.034	0.1
Al	0.001%	300	400	200	300	0.013	0.01
Na	0.001%	840	830	460	420	24.36	21.06
K	0.001%	300	300	1200	10600	10.047	1.82

Table 5. Content of elements in moss, algae and water in the Euphrate River,* = ppb; the other elements are given ppm.

2.3 Conclusions

The present study focused on the environmental geochemistry samples of water, stream sediment, algae and moss from the Keban mine district. The average concentrations of Mo, Cu, U, Sb, Mg, Hg, Se ve SO_4, Sr, V, Ca, P, Ba, B, Na, and Sc in Euphrates River water collected from near the former mining district were found to be elevated relative to average element concentrations reported for river waters. The concentrations of Al, Cd, Cr, F, Fe, Mg, Mn, Na, Pb, Sb, SiO_2, Tl, and Zn in water samples collected near the mineralized locations were higher than the maximum permissible standards for drinking water and, thus, it was determined that these waters are not drinkable. Consequently, these waters, which are used for drinking and agricultural purposes in rural areas, may pose a serious risk to public health. The high concentrations of As, Mo, Se, Cu, Zn, and Pb in the Karamağara Stream and Euphrates River sediments and waters suggest that both mechanical and chemical dispersions are dominant in the area. The element concentrations in tissues of green and red algae collected from the Euphrates River bed were compared with those in moss collected from the mouth of the mine adit. The concentrations of Cu, Pb, Ni, Mn, Au, V, and Cr in *Bangia atropurpurea* were found to be higher than those in *Cladophora glomerata*, but Zn, Co, Fe, Cd, Hg, Se, Ga levels in *Cladophora glomerata* are much higher than those in *Bangia atropurpurea*. The Cu, Pb, Zn, Mn, Ag, Ni, Co, Fe, As, U, Sr, Cd, Ca, K, Mg, Ba, and Au contents of *Didymodon tophaceus (Brid.) Lisa* are higher than those of the algae. Total BAF values for *Bangia atropurpurea* are higher than for *Cladophora glomerata*, while the total BAF values of the moss are highest.

These geochemical results suggest that chemical leaching–precipitation reactions are taking place near the Keban mine slags. The spatial distributions of metals in the sampling media of the Keban mining district suggest that the causes of the contamination are primarily the previous mining activities and to a lesser extent natural weathering products of the various types of mineralized zones in the area.

3. Maden copper deposit

3.1 Sampling and analytical methods

The second study area is Maden Copper deposit in which the water and stream sediment geochemistry were studied (Figure 3; Figure 4). The water samples were collected from

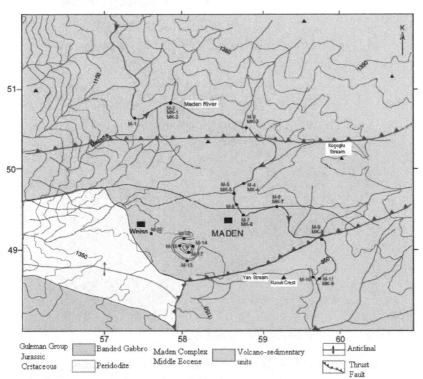

Fig. 3. Geology map of the Maden copper deposit and sediment samples location (MK: stream sediment samples, M: water samples; modificated Kırat et. al.2008; Kalender et. al. 2010).

Fig. 4. AMD waters in the Maden copper deposit.

M1- M11 (spring water samples; M12-M17 seepage water samples; M18-M24 waste water samples from copper flotation) and sediment samples from Maden Stream sediment then sediment samples were sieved in different size sediment fractions (range 850 to –500 µ; 500 to –180 µ; 180 to –106 µ; 106 to –75 µ; <75 µ) analyzed by ICP-OES and IC (for SO₄).

3.2 Results and discussion

The water samples taken from the vicinity of copper deposits within the scope of this study were determined for the metal content.

The waste waters from copper flotation (M18-M24) has increased over the metal content than the other samples. However, the mean values of analysis of water for all the sample points are given in Table 6 and these values are compared with drinking water standards. According to Table 6 Al, Ca, Fe, Mg, Mn and Zn values are higher than drinking water contents. Kalender & Bölücek (2007) indicated that metal contents are higher than maximum contaminat levels for drinking water and quite high enough to be dangerous to agriculture and stock raising and, thus to human health.

Elements	Max.Conta-minant Level (mgl-1)	Water samples	
		Mean (mgl-1)	Max. (mgl-1)
Al*	0.20	20.90	53.125
Ca	10-100	125.26	478.832
Cd	0.003	0.042	0.156
Co	1	10.42	21.546
Cu	1	107	314.962
Fe	0.2	9.65	22.106
Mg	125	486.47	491.564
Mn**	0.05	7.72	27.253
Ni	0.02	0.90	3.513
Pb	0.0015	0.0092	0.0388
Se	0.01	0.058	0.075
Zn	5	51.02	43.304
SO_4***	400	162	13224

Table 6. Drinking water maximum contamination level, (*: EEC, **: USEPA, ***: EQS fresh water, the others from WHO, standard values taken from Pais & Jones, 2000; Siegel, 2002). n=24.

Within the scope of this study, Cu, Fe, Mn, Zn, Ni and SO_4 values of two seepage water in location no M15-M16 which had the highest metal concentration were analyzed before and after sulfate-reducing bacteria experiment (Figure 3 and 4; Kalender et.al. 2010).

Table 7 presents Cu, Fe, Mn, Zn, Ni and SO_4 values analyzed before and after adding bacteria to AMD water samples (M15, M16). It was found that in M15 and M16 locations, Cu values decreased from 151 μgl-1 to average 5.82 μgl-1, from 1447 μgl-1 to average 29.1 μgl-1; Fe values decreased from 132.2 μgl-1to average 2.43 μgl-1, from 517 μgl-1to average 9.92 μgl-1; Mn values decreased from 75.7 μgl-1to average 19.33 μgl-1, from 1695 μgl-1to average 449.46 μgl-1; Zn values decreased from 34.4 μgl-1to average 0.45 μgl-1, from 3212 μgl-1to average 40.22; Ni values decreased from 4.3 μgl-1to average <0.1, from 70.4 μgl-1to average 4.49 μgl-1; SO_4 values decreased from 1417.8 μgl-1to average 788.38 μgl-1and from 1719.2 to average 925.22 respectively.

			Before Experiment			
Sample code	Cu	Fe	Mn	Zn	Ni	SO$_4$
M15	151.1	132.2	75.7	34.4	4.3	1417.8
M16	1447	517	1695	3212	70.4	1719.2
Detection Limits	0.1	10	0.05	0.5	0.2	0.1mgL^{-1}
			After Experiment			
	Cu	Fe	Mn	Zn	Ni	SO$_4$
M15-1-0.2	1.81	2.30	18.92	0.34	0.55	784.2
M15-2-0.2	1.70	2.40	19.23	0.37	<0.1	794.7
M15-1-2	2.63	2.60	19.45	0.42	<0.1	778.4
M15-2-2	2.7	2.72	19.98	0.48	<0.1	786.2
M15-1-20	13.51	2.25	18.9	0.52	<0.1	785.3
M15-2-20	12.6	2.28	19.5	0.56	<0.1	798.5
M16-1-0.2	36.18	9.07	423.5	30.2	3.6	924.4
M16-2-0.2	35.20	10.6	426.2	36.5	4.7	919.3
M16-1-2	23.12	10.7	452.5	40.5	4.5	920.5
M16-2-2	25.6	10.96	460.2	43.80	5.6	925.6
M16-1-20	26.7	8.87	462.2	44.66	4.33	932.5
M16-2-20	27.9	9.32	472.2	45.67	4.2	929.3

Table 7. pH values and metal contents (μgl^{-1}) of acidic seepage waters in two different sample collection locations before and after bacteria experiment M1:, 1: G20, 0.2 ml; M1:, 2: NO132, 0.2 ml; M1, 1: G20, 2ml; M1; 2: NO132, 2ml; : M1; 1: G20; 20 ml; M1, 2: NO132, 20ml; M2:, 1: G20, 0.2 ml; M2:, 2: NO132, 0.2 ml; :M2, 1: G20, 2ml; M2; 2: NO132, 2ml; M2; 1: G20; 20 ml; M2, 2: NO132, 20ml.

According to Table 8, alkalinity values of M15 and M16 locations were found to be 164 and 143 ppm before bacteria cultivation. EC (electrical conductivity) value of the mentioned acidic seepage waters was found to be 2056-2200 μS/cm. These values must have resulted from high metal content in the solution. *Desulphovibrio desulfurican* (G 20 - NO 132) sulfate-reducing bacteria used in the tests in the present study can covert the sulfate produced from the Maden main deposit (SO$_4$$^{-2}$) in the water which is rich in SO$_4$ into sulfur (S^{-2}). The bacteria can form bicarbonate (HCO$_3$) in the presence of an organic carbon source using sulfate as an electron donor under anoxic and reducing conditions (Bechard et al.,1995; Dalsgaard & Bak, 1994; Fyson et al.,1995).

$$SO_4{}^{2-} + 2CH_2O^* - Bacteria \rightarrow H_2S + 2HCO_3{}^- \tag{1}$$

CH$_2$O*= Ordinary organic molecule (lactate or acetate)

During the process of sulfate-reducing, firstly HS$^-$ is formed. HS$^-$ forms hydrogen sulfur (H$_2$S) by reacting with free hydrogen ion (1). Then, hydrogen sulfur reacts with metals and form insoluble metal complexes and thus metals removal occurs. Produced bicarbonate provides higher alkalinity and pH in the medium when compared to the initial values. Electrical conductivity (EC, μS/cm) which is lower than the initial value, suggests decreased metal concentration in the solution (Table 8). Sulfate-reducing *Desulphovibrio desulfurican* is not active in conditions with a pH lower than 5 (Egiebor & Oni, 2007). For this reason,

Sample code	Sulphate reducing bacteria	ml	pH	EC μS/cm	Alk ppm
Before experiment					
M15			5.1	2056	164
M16			2.5	2200	143
After experiment					
	G20	0.2	8.7		
	G20	2	7.8	2000	250
	G20	20	6.2		
M15	NO-132	0.2	8.2		
	NO-132	2	7.8	2009	238
	NO-132	20	7.2		
	G20	0.2	8.6		
	G20	2	8.5	2083	279
M16	G20	20	8.2		
	NO-132	0.2	7.8		
	NO-132	0.1	7.6	2030	252
	NO-132	1	7.2		

Table 8. The effects of two types of bacteria cultivated to different media on pH, EC and alkalinity values.

reaction was realized in newly created neutralization conditions. Since suitable pH value for the survival of the bacteria was 7.8, a pH value of the solution was set to 7.8. After the incubation of the bacterium, pH values increased from 5.1 to 8.7 in M15 location and from 2.5 to 8.6 in M16 locations (Table 7; 8). It was found that the bacterium 0.8 increased pH value.

Reduction of manganese and iron can make significant contributions to neutralization process.

It was observed that Fe value increased approximately 57 times while Mn values decreased 4 times. Heterotrophic bacteria like *Desulfovibrio* analyzed in the study can directly reduce iron and manganese by using iron as the last electron acceptor in anaerobic conditions. When Fe^{3+} is reduced to Fe^{2+}, removal of iron from AMD becomes easier. Fe^{2+} reacts with the sulfide formed after sulfate reduction and as a result, contributes to removal of iron and increase of alkalinity (Akcıl & Koldaş, 2006). The fact that Zn value decreased approximately 100 times suggests that like Fe, Zn also reacted with sulfate. When hydrogen sulfur gas (H_2S) which is removed from Maden waste sites during sulfate reduction, permanent alkalinity values can be determined. Ni values in sulfate containing solution varied from 5.6 μg/l to below detection limit.

Element distributions in different sediment size factions are presented in Table 9. The fact that the elements in the study area were in the form of primary sulfides indicates that chemical degradation is low. Aqua regia is a quite good solvent for these types of samples (Allcott & Latin, 1978; Rubeska et al. 1987). This mixture can also dissolve colloidal metals, oxide sediments and minerals and many elements can be analyzed by dissolving of only one sample. As the metal distribution dissolved with this method decrease in current flow

Elements	Detection Limits	-20+35	Size fractions -35+80	-80+140	-140+200	-200
As	0.1	31.52	47.84	73.28	63.1	59.64
B	1	6	6	6.6	6.6	5.4
Ba	0.5	55.26	69.86	82.9	93.08	110.36
Bi	0.02	0.428	0.638	1.146	1.142	1.324
Cd	0.01	0.376	0.494	0.988	1.054	1.134
Co	0.1	292.28	302.08	310.8	253.8	297.42
Cu	0.01	1419.35	2153.28	3877.24	4261.61	5198.03
Cr	0.5	227.48	237.38	216.54	198.82	179.98
Ga	0.1	6.42	6.44	6.22	6.24	6
La	0.5	6.44	6.56	6.48	6.54	7.3
Mn	1	1068	1170	1266.2	1292.4	1642.8
Mo	0.01	7.34	8.352	8.878	7.864	7.664
Ni	0.1	151.46	164.54	198.64	205.16	195.26
Pb	0.01	82.04	126.28	270.37	284.16	318.98
Sb	0.02	1.87	3.13	7.064	7.306	6.878
Sc	0.1	8.3	8.54	8.5	8.56	8.2
Se	0.1	2.06	2.9	6.86	7.48	8.08
Sr	0.5	40.94	41.4	47.06	54.9	68.66
Te	0.02	0.21	0.31	0.65	0.72	0.75
Th	0.1	0.92	0.92	0.98	1	1.04
Tl	0.02	0.19	0.25	0.43	0.45	0.51
U	0.05	0.32	0.36	0.68	0.74	0.62
V	2	105.2	120	111.2	99	92.6
Zn	0.1	832.64	760.98	533.4	472.74	485.44
ppb						
Ag	2	301.4	477	1071.2	1133	1441.8
Au	0.2	27.56	64	960.3	124.22	283.3
Hg	0.1	40	64	170.2	198.4	280
%						
Al	0.01	2.06	2.14	2.29	2.39	2.33
Ca	0.01	1.56	1.52	2.03	2.65	3.68
Fe	0.01	8.76	9.37	9.15	7.52	6.96
K	0.01	0.08	0.08	0.10	0.11	0.11
Mg	0.01	2.19	2.28	2.74	2.71	2.35
Na	0.001	0.01	0.01	0.02	0.02	0.02
P	0.001	0.04	0.04	0.04	0.05	0.05
S	0.02	0.32	0.55	2.18	1.28	1.15
Ti	0.01	0.12	0.13	0.12	0.12	0.11

Table 9. Stream sediment element concentrations in different size fractions along the Maden River sediment size fraction=mesh; by Aqua regia ($3HCl+HNO_3$) digestion method.

direction, according to Rose et al. (1979) it suggests that mechanical distribution is effective in the region. For this reason, it can be stated that physical processes are more effective than chemical processes in distribution of element in stream sand. Higher enrichment of As, Co, Cr, Mo, Ni, Sb, Zn and particularly Au values in -80 mesh sediment grain size when compared to -200 mesh grain size suggest that these elements were enriched by attaching on silica, that this dissolution method is not effective in dissolving silica and particularly in releasing Au.

3.3 Conclusion

Metal concentration is high in the vicinity of Maden copper deposit both in surface and ground waters and in stream sand samples. For this reason, two locations which particularly have high metal contents were determined and the change in sulfate was tried to be observed through anaerobic purification method. However, the decrease in Cu, Fe, Ni, Mn and Zn values along with sulfate after anaerobic purification experiments indicated that bacteria can decrease the concentrations in water by precipitation of metals.

As a result of analysis indicate that water samples are above the standards of fresh water and it is also enriched in some toxic cations and anion such as Cu, Fe, Mn, Zn, Ni and SO_4. Cu 151.1-1447 mgl^{-1} (EQS fresh water max contamination level (mcl) 0,16 mgl^{-1}), Fe 132.2-517 mgl^{-1} (mcl 1 mgl^{-1}), Mn 75.7-1695 mgl^{-1} (mcl 0.30), Zn 34.4-3212 mgl^{-1} (mcl 0,125), Ni 4.3-1447 mgl^{-1} (mcl 0.20 mgl^{-1}). SO_4 values reach to 1417.8 mgl^{-1} and 1719.2 mgl^{-1} (mcl 400 mgl^{-1}) in different location in acide mine drainage waters. That is why the sulphate reducing bacteria experience was done in these water samples *Desulfovibrio desulfuricans* can initiate some metal precipitation in the parallel to indirect sulfide-mediated precipitation. Initial and final metals and sulphate concentration before and after experiments, were measured and finally evaluated influence of bacteria in different ratio (0.2; 2; 20 ml) in AMD waters at two different sample locations (M15 and M16) . The experimental results show that, Cu, Fe, Mn, Zn, Ni and SO_4 values have been decreased by using sulphate reducing bacteria, Cu values, range of 96 to 98 % ; Fe values, 98 %; Mn values, range of 73 to 74%; Zn values, 98 %; Ni values, range of 93.62 to 97.67% ratio and SO_4 values, range of 44.36 to 46.18% ratio, respectively.

Initial, metals are dissolved due to <pH in AMD waters, but after bacteria experiments metals are precipitated as complex compound due to >pH (>7 at the neutralize pH condition) and high alkalinity. Consequently, in this study bacteria experiments obseved that SRB (sulphate reducing bacteria) affected on both precipitation of the metals and the sulphate reducing in the AMD waters from the Maden copper deposit. These data will be evaluated including in literature and investigated impact on environment pollution.

In stream sediment samples reach to high concentrations Zn values –500 µ, Fe values –180 µ, As values –106 µ; Cu, Pb and Cd values <75 µ in size fraction. These values were correlated with average of diabase according to NIST and observed to high concentration of the Cu, Pb, Zn, As, Cd and Fe values. It was found that metal concentrations in stream sand samples increased mostly by mechanical transportation, however metal content in various water samples increased chemically in high reducing conditions.

4. Alacakaya chromite deposit

Thirth study area is Alacakaya alpine type chromite deposit in which the water geochemistry was studied (Figure1).

4.1 Sampling and material method

The water samples were collected from natural springs and ground water in the chromitite deposit area. The samples were taken using a hand pump and water sampler and filtered using 0.5 μm membrane filter paper (4.5 cm diameter); subsequently, 5 ml of prepared 65% concentrated HNO_3 solution were added to 250 ml of each water sample and the samples were stored in the laboratory at 4°C until analysed. Analysis were done in Kanada ACME Labs. by ICP-OES method. Detection limits and the results of the water samples analyses are given in Table 10.

4.2 Result and discussion

There are many waste place in chromite mining area (Fig. 5). The waste is an important factor on ground water contamination by leaching. The second effect is surface waters seepage to the deept and after mixing to ground waters (Fig. 6A). Figure 6B shows serpentine formation in the study area. The spring and ground water element contents in the Alacakaya chromite mining area are showed in Table 10.   When element contents are correlated to natural ground water chemical composition, Al, Fe and Mn values are high. Maximum Mg, Al and Mn values are 1912.706 mgl⁻¹, 672.72 mgl⁻¹ and 125.56 mgl⁻¹in groundwater samples in the Guleman Ophiolites, Upper Cretaceous. According to Rose et al. (1979) natural groundwater element contents were pointed out as Mg 7000 mgl⁻¹, Al 10 mgl⁻¹ and Mn 15 mgl⁻¹. However, the same element contents are correlated to drinking water maximum contamination levels that all metal contents are above drinking water maximum contamination levels (Table 10). The average concentration of Cr in the basaltic and ultramafic rocks the average concentration is 200 ppm. The average concentration of Mg in basaltic rocks the average concentration is 45000 ppm. The average concentration of Ni in the basaltic rocks is 150 ppm (Krouskopf, 1979). The results of rock samples analysis are given in Table 11.

Fig. 5. Waste in the Alacakaya Cr deposit.

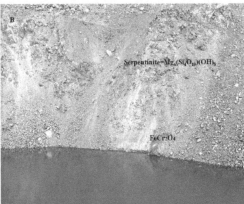

Fig. 6. Seepage waters from surface to deep (A) in the gallery; serpentinite and chromitite (B).

Element	Mean	Max	Min	Dedection Limits	Ground-water	Max. allowed values	Drinking water (MCL)
Al	75.576	672.72	0.005	1	10	50a	0.20
As	0.003	0.002	0.001	0.5	2	50a	-
B	0.1	0.31	0.03	20	10	-	-
Ba	0.01	0.018	0.003	0.05	20	50b	-
Ca	139.14	463.78	13.47	50	5000	-	10-100
Cd	0.03	0.16	0.0005	0.05	0.03	5b	0.005
Co	5.6	39.35	0.001	0.02	0.1	-	1
Cu	74	447.75	0.002	0.1	3	1000a.b	1
Cr	0.05	0.23	0.001	0.5	1	50a	0.1
Dy	0.02	0.18	0.001	<0.01	-	-	-
Fe	31.83	302.03	0.1	10	100	300a.b	0.3
Mg	252.88	1912.71	2.88	50	7000	50000b	50**
Mn	14.8	125.56	0.003	0.05	15	50b	0.05
Na	7.4	21.26	1.7	50	30000	20000b	20
Nd	0.07	0.65	0.0005	<0.01	0.08*	-	-
Ni	0.95	6.3	0.002	0.50		-	0.02
P	1.2	11.12	0.02	20	20	-	-
Pb	0.01	0.06	0.001	10	3	15b	0.015
Rb	0.002	0.013	0.003	0.01	1	-	-
S	0.6	4408	11	1000	30000	-	-
Y	0.09	0.85	0.004	<0.01	0.06*	-	-
Zn	10.18	60.48	0.03	0.50	20	5000a.b	5

Table 10. Element concentrations of the spring and ground water samples (in mgL^{-1}); element contents of natural ground water (in mgl^{-1}) from Rose et al. (1979); *= Jenssen & Verweij 2003; **= EEC; a- Ley 18284 (1969) Codigo Alimentario Argentino; Modificaciones (1988) and (1994); b- USEPA 1980, 2000. MCL= maximum contamination level; n= 10.

Major oxide (%) and minor element (ppm)	Harzburgite	Dunite	Pyroxenite	Detection Limits
SiO_2	44.56	43.3	49.56	0.01
Al_2O_3	0.63	0.46	6.33	0.01
Fe_2O_3	9.35	9.1	5.4	0.04
MgO	42.85	43.4	17.46	0.01
CaO	0.69	0.59	18.28	0.01
Na_2O	0.04	0.02	0.17	0.01
K_2O	<0.04	<0.04	<0.04	0.01
TiO_2	<0.01	<0.01	<0.01	0.01
P_2O_5	<0.01	<0.01	<0.01	0.01
MnO	0.12	0.11	0.12	0.01
Cr_2O_3	1.14	0.51	0.41	0.002
Co	110.9	11.5	46.2	0.001
Ni	2634	2682	231	20*
Cu	14.5	25.4	32.7	0.0001
Zn	36	58	42	0.0001
Rb	0.6	<0.5	<0.5	0.5*
Nd	<0.4	<0.4	0.7	0.4*
Dy	<0.05	<0.05	0.65	0.005*
Y	0.2	0.3	6.6	0.01*

Table 11. Major oxide and minor element concentrations of harzburgite, dunite and pyroxenite from the Alacakaya chromitite deposit area from Başpınar (2006), Detection limits in %; * = ppm.

The high metal concentration in both spring and ground water samples indicated that source of contamination is lithological units (harzburgite, dunite, pyroxenite and alteration produce serpentinite) due to surface erosion, leaching and tectonism.

Talc is an alteration mineral. Altering serpentine ($3MgO \cdot 2SiO_2 \cdot 2H_2O$) can also form talc. Serpentine commonly contains chrysotile asbestos, a carcinogen listed by the EPA. That is why; groundwater geochemistry was investigated in the Alacakaya chromite deposit area (Kalender, 2009).

4.3 Conclusion

These studies show that the major and minor elements may be useful geochemical tracers but they have dramatically effect on environmental pollution especially in mining area. The present studies have demonstrated that the distribution and behavior of toxic and non toxic element content in the different media (water, sediment, and plants) may be bonded to high concentrations in geological units.

This study focuses on environmental effection of the metal pollution in the different origin ore deposits.

Environmental Impact and Drainage Geochemistry of the Abandoned Keban Ag, Pb, Zn Deposit, Working Maden
Cu Deposit and Alpine Type Cr Deposit in the Eastern Anatolia, Turkey

103

5. Anklowledgement

Keban Ag-Pb and Zn and Maden Cu deposit studies were financially supported by Firat University, Grant Number 628 and 1506, respectively.

6. References

Akcıl, A. & Koldaş, S. (2006). Acid Mine Drainage (AMD): Causes, treatment and case studies. *Journal of Cleaner Production*, Vol.14, pp. 1139-1145, ISSN 0959-6526

Akıncı, O., Acar, E. & Tüfekçi, S. (1977). A study of Plan Keban Pb–Zn ore, *General Directorate of Mineral Research and Exploration: Report*, pp. 68-76, ISSN 0026-4563

Aksın, M., Çetin, A. K. & Yıldırım, V. (1999). The Algae of Keban Stream (Elazığ, Turkey), *Fırat University, Journal of Science*, Vol. 11, pp.59–65, ISSN 1308-9072 (in Turkish).

Allcott, G.H. & Latin, H.W. (1978). Tabulation of Geochemical Data Furnished by 109 Laboratories for Six Explorations Reference Samples. *U.S. Geological Survey, Open File Report*, pp. 78-163, ISSN 0196-1497

Balçık, A. (1979). Mineralizations of Keban Nalliziyaret and Karamağara Dere (Bamas), *General Directorate of Mineral Research and Exploration: Report*, pp. 23, ISSN 0026-4563

Başpınar, G., (2006). Geochemistry and Contents of Platinum Group Elements of Guleman Chromite Deposit, Elazig, Turkey. *Firat University, Science Institute, Master Thesis*, pp. 125. Elazig, Turkey (Unpublished).

Béchard, G., Mc Cready, R. G. L., Koren, D.W. & Rajan, S. (1995). Microbial Treatment of Acid Mine Drainage at Halifax International Airport Sudbury, Ontario. Canada, In: Proceedings of Sudbury'95: *Mining and the Environment Conference Proceedings*, Sudbury, Ontario. 28 May – 1 June, Vol. 2, pp. 545-554. ISSN 1319-8025

Bölucek, C. (2002). A Stream Sediment Geochemical Orientation Study in Derince (Keban–Elazığ) Vicinity, *Bulletin of Earth Sciences, Application and Research Centre of Hacettepe University*, Vol. 25, pp. 51–63, ISSN 1301-2894 (in Turkish).

Bölücek, C., (2007). Environmental Contamination in The Keban Mining District, Eastern Turkey, *AJSE*, Vol.32/1A, pp.3-18, ISSN 1319-8025

Bölücek, C. & Kalender, L, (2009). Drainage Sediment Geochemical Orientation Study Under Semi-Arid Coditions, Keban, Eastern, Turkey. *AJSE*, Vol. 34, 1A, pp. 91-102, ISSN 1319-8025

Çalık, A. (1998). Keban Plutonits; its Relation to Mineralogy, Petrogenesis and Wall Rocks, *Ph. D. Thesis, University of Istanbul*, pp. 18, 1 Istanbul,Turkey, (unpublished).

Çelebi , H. & Hanelçi, Ş. (1998).Geochemische und Geostatistische Untersuchungen an Mn-Erzen des Lagerstaettendistriktes Keban, Elazığ/Osttürkei, *Geologisches Jahrbuch*, Hannover, Vol.108, pp. 3–33, ISSN0016-7851

Canada MAC, *Guidelines for Canadian Drinking Water Quality*, 2001.

Dalsgaard, T. & Bak, F. (1994). Nitrate Reduction in a Sulfate-Reducing Bacterium, Desulfovibrio desulfuricans, Isolated from Rice Paddy Soil. Sulfide Inhibition, Kinetics, and Regulation. *Applied and Environmental Microbiology*, Vol. 60 (1), pp.291-297, ISSN00992240

Egiebor, N.O. & Oni, B. (2007). Acid Rock Drainage Formation and Treatment: A Review. *Asia-Pasific Journal of Chemical Engineering*, Vol. 2, pp. 47-62, ISSN 1932-2135

Entwisle, T. J. (1989). Phenology of the Cladophora-Stigeoclonium Community in to Urban Creeks of Melbourne, *Australian Journal of Marine and Freshwater Research,* Vol. 40, pp. 471–489, ISSN 0067-1940

Environmental Quality Standards (EQS) are Annual Average Concentrations with Maximum Allowable Concentrations in brackets. Esdat Environmental Database Management Software, www.esdat.net

Erdoğan, B. (1977). Geology, Geochemistry and Genesis of the Sulphide Deposits of the Ergani-Maden Region, SE Turkey. Ph.D. Thesis, The University of New Burnswick, p. 289 (unpublished).

Erdoğan B. (1982). Geology and volcanic rocks of the ophiolitic zone in the Maden area. *Geological Bulletin of Turkey, Vol.* 25, pp.49-60, ISSN 1016-9164 (in Turkish).

Europium Economy Community (EEC), (1992). Standard Statistical Classification of Surface Freshwater Quality for the Maintenance of Aquatic Life. CES/733, 13 April 1992. United Nations, Economic and Social Council.

Fyson, A. Kalin. M. & Smith, M.P. (1995). Nickel and Arsenic Removal from Mine Wastewater by Muskeg sediments. *Proceedings of the 11th Annual General BIOMINET Meeting, Ottawa, Ontario,* January 16, pp. 103-118.

Janssen R.P.T. & Verweij, W. (2003). Geochemistry of some rare earth elements in groundwater, Vierlingsbeek, The Netherlands. *Water Research, Vol.* 37, pp. 1320–1350, ISSN: 0043-1354

Kalender, L. (2000). Geology, Origin and Economic Importance of the Copper Mineralization of Keban (Elazig), East Euphrates Kebandere Area, *Ph.D. Thesis, Firat University, pp. 110, Elazığ, Turkey.* (in Turkish).

Kalender L. & Hanelçi, Ş. (2001). Mineralogical and Petrographical Features of Nallıziyaret Tepe (Keban–Elazig) Copper Mineralization, *Istanbul University Earth Sciences,* Vol.14, pp. 51–60, ISSN 1016-9806 (in Turkish).

Kalender L. & Hanelçi, Ş. (2001). Mineralogical and Geochemical Features of Au, Ag, Pb and Zn Mineralization in Keban (Elazığ) Wastes, *Geological Bulletin of Turkey, Vol.* 44, pp.91–104, ISSN 1016-9164 (in Turkish).

Kalender L. &Hanelçi, Ş. (2002). General Features of Copper Mineralization Nallıziyaret Tepe (Keban–Elazığ): An Approach to its Genesis, *Geosound,* Vol. 40/41, pp.133–149, ISSN 1019-1003 (in Turkish).

Kalender, L. & Bölücek, C. (2004). Major And Trace Elemet Contamination of Groundwaters, Stream Sédiments and Plants of the Abandoned Mines in Keban District (Elazığ) of Eastern Anatolia, Turkey. *57. Geological Congress of Turkey.* 08-12 March, pp.187-188.

Kalender, L. & Bölücek, C., (2007). Environmental Impact and Drainage Geochemistry in the Vicinity of the Harput Pb-Zn-Cu Veins, Elazığ, SE Turkey, *Turkish Journal of Earth Sciences,* Vol.16, pp. 241-255, ISSN 1300-0985

Kalender, L. (2009). Impacts on Environmental of Ore Deposits in Elazig Area. Ministry of Environment and Forest, Environmental Legislation Remediation and Applied Conference, (oral presentation), TAIEX, 25 March 2009, Elazığ, Turkey.

Kalender, L., Kırbağ, S., Kırat, C. & Bölücek, C., (2010). The Sulfate Reducıng Bacterıum Experıments In AMD Waters of The Maden (Elazığ) Copper Deposıt, Firat University, *Journal of Engineering,* Vol. 22/2, pp. 197-204, ISSN 1308-9072

Kalender, L., Kırbağ, S., Kırat, C. & Bölücek, C., (2010). The Sulfate Reducing Bacterium Experıments in AMD Waters of the Maden (Elazığ) Copper Deposit, Seventh

International Symposium On Eastern Mediterranean Geology, 18-22 October 2010, pp. 130, Adana /Turkey.

Kelley, D.L., Hall, G.E.M., Closs, L.G., Hamilton, I.C. & Mc Ewen, R.M. (2003). The Use of Partial Extraction Geochemistry for Copper Exploration in Northern Chile. *Geochemistry: Exploration, Environment, Analysis*, Vol. 3, pp. 85–104, ISSN 1467-7873

Kineş, T. (1969). The Geology and Ore Mineralization of the Keban Area, Eastern Turkey, *Ph.D. Thesis, Istanbul University*, pp.213, Istanbul, Turkey, (unpublished).

Kipman, E. (1976). Geology and Petrology of Keban, *Ph.D. Thesis, Istanbul University*, pp.189, Istanbul, Turkey, (unpublished).

Kırat, G., Bölücek, C., Kalender, L., (2008). Distribution of Cu, Pb, Zn, As Cd and Fe in stream sediments around of Maden copper deposit. *Geosound/Yerbilimleri*, Vol.53, pp. 203-217, ISSN 1019-1003.

Kishler J. & Taft, C.E. (1970). *Bangia* atropurpurea (Roth) in Western Lake Erie, *Ohio Journal of Science*, Vol. 70, pp. 56–57, ISSN 0050-0950

Köksoy, M. (1975). Geochemical Leakage Anomalies in the Vicinity of Keban Mine, *Bulletin of Geological Society of Turkey*, Vol. 18, pp. 131–138, ISSN 1016-9164 (in Turkish).

Krauskopf, K. B. (1979). Introduction to Geochemistry. Tokyo: McGraw-Hill Kogakusha, Ltd., ISBN 0070358206

Kumbasar, I., (1964). Petrographic and Metallogenic Features Mineralisations in Keban Vicinity, *Ph. D. Thesis, Istanbul Technical University*, pp. 157, Turkey, (unpublished).

Ley 18284. Código Alimentario Argentino (1969). Sancionada y promulgada el 18/7/69. B.O. 28/7/69. Modificación 1988 y Modificación 1994, Buenos Aires, Argentina.

Özdemir, Z. & Sağıroğlu, A. (1998).The Study of Biogeochemical Anomalies for Fe along the Maden River (Maden-Elazig). *Geological Bulletein of Turkey*, Vol. 41/1, pp. 49-54, ISSN 1016-9164

Özdemir, Z. & Sağıroğlu, A. (2000). The Study of Biogeochemical Anomalies for Zn along the Maden River (Maden-Elazig). *Mersin University, Series of Essay*, Vol. 4, pp. 93-100, 0002-6417

Özkaya, İ. (1978). Stratigraphy of Ergani-Maden vicinity. *Geological Bulletein of Turkey*, Vol. 21, pp. 129-139, ISSN 1016-9164

Pais, I. & Jones J. B. Jr. (2000). *The Handbook of Trace Elements*, St. Lucie Press, Boca Raton, ISBN 0849314585, Florida,

Power, M. E. (1992). Benthic Turfs vs. Floating Mats of Algae in River Food Webs, *Oikos*, Vol.58, pp. 67–79, ISSN 0030-1299

Rose, A. W., Hawkens, H.E. & Webb, J.S. (1979). *Geochemistry in mineral exploration*. (2nd edition). Academic Press, ISBN 0-7718-8893-7, London.

Rubeska, I., Ebarvia, B., Macalalad, E., Ravis, D.& Roque, N. (1987). Multi Element Preconcentration by Solvent Extraction Compatibible with an Aquaregia Digestion for Geochemical Exploration Samples. *Analyst*, Vol. 112, pp. 27-29, ISSN 0003-2654

Schönborn, W. (1996). Algal Aufwuchs on Stones, with Particular Reference to the Cladophora–Dynamics in a Small Stream in Thuringia. (Germany): Production, Decomposition and Ecosystem Reorganizer, *Limnologica*, Vol. 26, pp. 375–383, ISSN 0075-9511

Seeliger, T.C., Pernicka, E., Wagner, G.A., Begeman, F., Schmitt-Strecker, S., Eibner, C., Öztunalı, Ö. & Baranyı, I., (1985). *Archo-metalurgische untersue-hungen ni Nord und*

Ostanatolien. 32. Jahbuch des Römisch. Germanischen Zentralmuseums, p.597-659, Mainz, Germany.

Siegel, F. R. (2002). *Environmental Geochemistry of Potentially Toxic Metals,* Springer-Verlag GmbH, pp. 218, ISBN 3 540 42030 4, Berlin and Heidelberg, Germany

Tızlak, E, (1991). Mining in the Keban - Ergani area (1780- 1850). *Ph.D. Thesis, Firat University, pp. 402, Elazığ, Turkey.* (in Turkish).

United State Environmental Protection Agency, (2002). *List of Drinking Water Contaminants and MCL.*

Whitton, B.A., John, D. M., Johnson, L. R. , Boulton, P. N. G. , Kelly, M. G. & Haworth, E.Y. (1998). Perspective on the Coded List of the Freshwater Algae of the British Isles. The *Science of the Total Environment,* Vol. 210/211, pp. 283–288, ISSN 0048-9697

World Health Organization, (1993). *Guidelines for Drinking-Water Quality,* Geneva: WHO.

Yılmaz, A., Ünlü, T. & Sayılı, S. (1992). An Approach to Mineralization of Pb–Zn in Keban (Elazığ), *Mineral Researchand Exploration Geology Review,* Vol. 114, pp. 47-70 (in Turkish). ISSN 0026-4563

Zisserman, A. (1969). Geological and Mining Study of Keban Madeni, Elazığ (Turkey), *Ph.D. Thesis, BRGM France.*

5

Lanthanides in Soils: X-Ray Determination, Spread in Background and Contaminated Soils in Russia

Yu. N. Vodyanitskii and A. T. Savichev

Department of Soil Science, Moscow State University; Geological Institute, RAS
Russia

1. Introduction

Rare elements in soils are poorly studied because of the problems in their determination (Perelomov, 2007). The rare earth elements include lanthanum (La) and its group of 14 elements. They are subdivided into two subgroups: the light cerium group of elements with atomic masses lower than 153 (La, Ce, Pr, Nd, Sm, and Eu) and the heavy yttrium group of elements with atomic masses higher than 153 (Y, Gd, Tb, Dy, Ho, Er, Tm, Yb, and Lu, with an exception for Y) (Tyler, 2004a).

The interest in rare earth elements in soils increased in the 1990s, when they found wide use as microfertilizers for some crops in China, which increased the yield and quality of grain (Evans, 1990; Wu, Guo, 1995; Zhu et al., 1995). Rare earth elements are accumulated in soils due to fertilization. Groundwater and plants can be contaminated in the regions enriched with soluble rare earth metals and in the soils fertilized with sewage sludge for long time (Zhu et al., 1995). Data are available on the strong contamination of coastal sediments with lanthanidecontaining industrial waste (Savichev, Vodyanitskii, 2009). Neither positive nor negative geochemical anomalies of natural or technogenic rare earth metals have been revealed in Russia until now. The contribution of rare earth metals to the main soilforming processes is not clear.

The content of lanthanides was compared to their soil clarkes. Questions arise in this connection. The yttrium clarke value is 40 mg/kg (Bowen, 1979). We shall take this value, although we consider it to be too high.

We analyzed the clarkes of the rare earth elements (La and Ce) according to the book by Kabata-Pendias and Pendias (1985), in which they were borrowed from three sources. Let us judge the validity of the published La and Ce clarkes according to the Ce : La ratio in the world soils. The clarkes proposed by Lawl et al. (according to Kabata-Pendias and Pendias, 1985) are the least probable of all, as they give a too low ratio of Ce : La = 29.5 : 29.5 = 1. On average, the Ce : La ratio = 2.2 for the soils in Japan; it is equal to 2.0, for the soils in China; and to 1.8, for the soils in Bryansk oblast of Russia. According to our data, this ratio is equal on average to 1.5 in the soils of the Kolyma depression, as well as in the alluvial fine earth of the small rivers and the Kama River in Perm (Savichev, Vodaynitskii, 2009). Bowen (1979) also suggests too low a ratio (Ce : La = 50 : 60 = 1.2). This ratio reaches a probable value only for the clarkes proposed by Yuri and Baikon: Ce : La = 49 : 34 = 1.4. These values of the lanthanum (34 mg/kg) and cerium (49 mg/kg) clarkes we shall use further. In sedimentary rocks, the clarkes of La and other lanthanides depend on the particle size distribution, because these elements are accumulated in the clay fraction <2 μm. As a result, their content is higher in argillites and clays (56 mg/kg) and lower in sandstones and carbonates (19 and 8 mg/kg, respectively) (Ivanov, 1997).

The clarkes of other studied lanthanides in the earth's crust are as follows (mg/kg): Pr, 9; Nd, 40; and Sm, 7 (Greenwood, Ernshaw, 1997). In soils, they are lower (mg/kg): Pr, 7.6; Nd, 19; and Sm, 4.5 (Dyatlova et al., 1988). The values of the soil clarkes are used to detect positive and negative geochemical anomalies. Unfortunately, there are no clear criterions for such identification. As a first approximation, we use the condition that a soil belongs to the territory of a strong positive anomaly at the double excess of the clarke ($X_s : X_{cl} > 2$).

2. Methods

The recent progress in the study of rare earth elements was related to the use of the expensive method of inductively coupled plasma mass spectroscopy (ICP MS) (Kashulina et al., 2007; Perelomov, 2007).

The expensive method of neutron-activation analysis using a nuclear reactor with the use of a gamma spectrometer allows determining the different dispersed elements in soils—Hf, La, Ce, Sm, Eu, Yb, Lu, Th, and U—even at low concentrations (Inisheva et al., 2007; Nikonov et al., 1999). However, the identification of Pr and Nd by this method is complicated because of the short lifetime of these elements (Ivanov et al., 1986).

X-ray fluorescence is the simplest and least expensive method for studying the heavy metals in soils (Savichev, Sorokin, 2000). The contents of rare (Zr and Nb) and rare earth (Y) elements are determined by this method. However, when a Mo, Rh, or Ag anode of the X-ray tube and a common voltage of 35–40 kV are used, other rare earth elements can be identified only from the L-lines. Their intensities are lower than those of the K-lines by several times, and the weak L-lines are overlapped by the strong K-lines of macroelements. Niobium with $Z = 41$ is the last element reliably identified by conventional X-ray fluorescence analysis, because the scattered lines of the anode material are located father in the energy spectrum, and the elements with atomic numbers higher than that of the anode material are not excited.

A radically different situation is observed when the X-ray radiometric modification of the energy dispersive X-ray fluorescence method is used, in which the sample is excited by a highenergy radioisotope source [241]Am rather than by the X-ray tube radiation. This source is the most suitable for this purpose. An advantage of this method is that the K-lines of the heavy metals are actively excited in this case and are not overlapped by the lines of macroelements. Other advantages are the low background radiation compared to X-ray tubes, the high stability, and the small size.

The aim of this part of work is to elaborate methods of X-ray radiometric determination of the first lanthanides (lanthanum and cerium) and barium and to determine the content of these elements in some soils.

Calibration diagrams were compiled with the use of standard samples of rocks and soils. There were 18 samples of magmatic rocks, including ultramafic (UMR), mafic (MR), acidic (AR), and alkaline (ALR) series (Khitrov, 1984). Thus, standard samples of gabbro (SGD-1A), granites (SG-1A and SG-2), trappean rock (ST-1), siltstone (SA-1), and others were analyzed. Standard soil samples included Kursk chernozem (SP-1), Moscow soddy-podzolic soil (SP-2), Caspian light chestnut soil (SP-3), krasnozem (SKR), and calcareous serozem (SSK). Detailed data on these standard samples are given in (Arnautov, 1987). It is important that the intensities of characteristic lines of studied elements do not depend on the mineralogical and textural specificity of the samples.

Conventional X-ray fluorescence and X-ray radiometric approaches were compared with the use of two energy dispersive X-ray fluorescence analyzers EX-6500 (Baird) and Tefa-6111 (Ortec). An EX-6500 analyzer had the following parameters: voltage 35 kV, current 400 μA, Rh-anode, Rh-filter, and storage time 400 s. The regime of a Tefa-6111 (Ortec) analyzer was as follows: voltage 30 kV, current 200 μA, Mo-anode, Mo-filter, and storage time 400 s. Samples were prepared for traditional working regimes via tableting. The main requirement for tablets was that they should produce saturated emission spectra, which was achieved upon their mass of about 3 g.

Instead of the traditional sample excitation by emanation from an X-ray tube, excitation by isotope source ^{241}Am with energy of 59 keV and activity of $3.7 \cdot 10^{10}$ s^{-1} was used for a Tefa-6111 analyzer. Preparation of the samples for the X-ray radiometric method was simple: powdered samples were placed in polyethylene dishes with 5-μm-thick Mylar bottoms. The mass of the powdered samples was about 8 g (exact weight of the sample is not required for this analysis). For an EX-6500 analyzer, the radioisotopic excitation was not efficient, because hard quanta of K-lines of the investigated elements penetrated through the detector and were only slightly slowed down by it.

Spectra of microelements in a standard sample of granite (SG-1A) obtained on an EX-6500 analyzer are given in Fig. 1. Upon a traditional X-ray fluorescence analysis, niobium is the last determined element. Elements with greater numbers are not excited, and scattered lines of the anode material are only situated farther on the energy spectrum.

The spectra from granite (SG-1A) and chernozem (SP-1) upon their excitation by the radioisotope source are shown in Fig. 2. The needed K-lines of Cs, Ba, La, Ce, and Nd are clearly seen. A comparison of the spectra shown in Figs. 1 and 2 indicated that the X-ray radiometric approach is useful for determining Cs–Nd series. Cesium is reliably determined, when its content exceeds 10 mg/kg, which is typical of the soils enriched by natural or anthropogenic Cs. Upon the X-ray radiometric excitement, K-lines of neodymium are overlapped by K-line (β components) of barium and are well seen in the samples with the low Ba content (in particular, in granites). In the soils with the Ba content above 400 mg/kg, the determination of neodymium is difficult. Thus, only Ba, La, and Ce are reliably determined in the soils.

We studied metrological characteristics of the new approach using a set of 30 standard samples of soils and rocks. The limit of element determination was 4 mg/kg at storage time of 800 s.

I, impulse/s

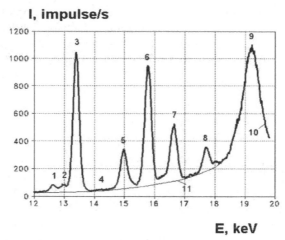

E, keV

Fig. 1. Part of the SSC SG-1A spectrum from Pb L_β to Zr K_β (EX-6500 analyzer, X-ray fluorescence method). Position of characteristic lines on the curve: (1) Pb L_β, (2) Th L_α, (3) Rb K_α, (4) Sr K_α, (5) Y K_α + Rb K_β, (6) Zr K_α, (7) Nb K_α + Y K_β, (8) Zr K_β, (9) incoherent scattering Rh K_α (material of anode), (10) distribution given by multichannel analyzer, (11) approximation of background radiation.

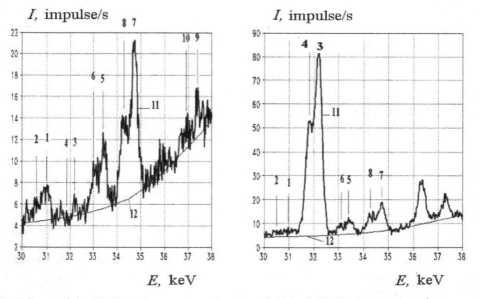

Fig. 2. Parts of the SSC SG-1A spectrum of granite (left) and SSC SP-1 spectrum of chernozem (right) from Cs K_α to Nd K_α (TEFA-6111 analyzer; X-ray radiometric method). Position of characteristic lines: (1) Cs $K_{\alpha1}$, (2) Cs $K_{\alpha2}$, (3) Ba $K_{\alpha1}$, (4) Ba $K_{\alpha2}$, (5) La $K_{\alpha1}$, (6) La $K_{\alpha2}$, (7) Ce $K_{\alpha1}$, (8) Ce $K_{\alpha2}$, (9) Nd $K_{\alpha1}$, (10) Nd $K_{\alpha2}$, (11) distribution given by multichannel analyzer, and (12) approximation of background radiation.

The error of the new analytic approach should be known. Absolute ΔC or relative $\Delta C/C$ deviation from the mean is usually used for error assessments. The advantage of the relative deviation is that it gives different estimates of the error upon high and low concentrations of determined elements. A multiplier $1/C$ is a statistical weight of an absolute error. However, relative deviation also depends on element concentrations changing within a wide range of values. The requirements of the Scientific Council on analytic methods (Berenshtein et al., 1979) are met upon the specification of element concentrations into small ranges and with the determination of the relative deviation in each range. The lower an element concentration, the higher the relative deviation. In order to obtain a more consistent estimate of the error working in a wide range of element concentrations, a different statistical weight should be chosen. Multiplier $1/C^{\frac{1}{2}}$ is the best statistical weight. The error measure equal to $\Delta C/C^{\frac{1}{2}}$ (K-factor) should be used (Savichev, Fogel'son, 1987). The K-factor makes it possible to describe the error within a wide range of concentrations without its subdivision into smaller ranges. In addition, the structure of K-factor agrees with the fact that the detection of X-ray roentgen spectra is based on the counting of impulses, which is a Poisson process, for which the variance correlates with the square root of the mean.

The accuracy of determination corresponds to the third category (according to the requirements of the Scientific Council on analytic methods) at the lower limit of determinable concentrations (8 mg/kg), to the second category at concentrations about 100 mg/kg, and to the first category at the concentrations about 500 mg/kg. Metrological data on the Ba, La and Ce elements are given in Table 1.

Element	Range of measured concentrations, mg/kg	K-factor
Ba	10–2000	1.28
La	10–150	1.02
Ce	10–250	1.07

Table 1. Metrological data on the X-ray radiometric analysis of microelements in soils.

The detection of Pr, Nd, and Sm faces some problems. Their determination is complicated by the fact that the $K_{\alpha1,2}$ lines of Pr and Nd are overlapped by the Ba $K_{\beta1,2}$ and La $K_{\beta1}$ lines. The determination of Pr is also complicated by its low content in soils. The problems of the Sm identification are the same as for Pr and Nd.

The general view of the spectra for the region of elements from Ba to Sm is shown in Fig. 3A for the SGD-1A gabbro–essexite reference standard. An especially strong overlapping is observed in the energy region of 35–38 keV, which appreciably deforms the true intensities of the Pr and Nd spectral lines. A detailed view of this spectral region is shown in Fig. 3B.

The main objective in the determination of the true line intensities of Pr and Nd is to correctly simulate the disturbing Ba $K_{\beta1,2}$ and La $K_{\beta1}$ lines. For this purpose, an adequate model should be selected for the spectral line contour. There are two main approaches. The first is an experimental approach in which the line contours of pure elements are used as model spectral lines. The second is an analytical approach in which the model line contour is specified by a mathematical function (in our case, a Gaussian curve). However, both approaches presume the intensity ratio of the α and β components to be stable. The experimental ratio between the Ba $K_{\alpha1}$ and Ba $K_{\beta1}$ line intensities is 5:1. However, in the spectra of real samples, this ratio was found to be slightly lower (by 20%) and varied (by about 10%) among the samples; i.e., this ratio is subjected to matrix effects, which should be taken into consideration in the determination of the true intensities of the Pr and Nd lines. Therefore, the Gaussian model was used as a model of the spectral line contours, and the ratio between the α and β components was selected for each sample on the basis of the best agreement between the top segment of the Ba $K_{\beta1}$ line and the experimental spectrum, as is shown in Fig. 3B. An adjustment of the model contour on the energy scale was also performed: a parallel shift within an arbitrary number of the analyzer channels (for the energy resolution, 10 eV/channel).

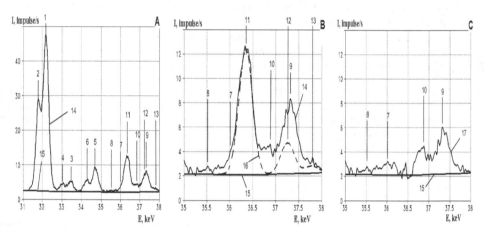

Fig. 3. Spectral segment corresponding to the K-lines of Ba, La, Ce, Pr, and Nd for the SGD-1A (gabbro–essexite) reference standard: (A) general spectrum view; (B) detailed interpretation of the spectrum for Pr and Nd with the simulation of the K_β components of the Ba and La lines; (C) the spectrum after the subtraction of the K_β components. The spectral lines: (1) Ba $K_{\alpha1}$; (2) Ba $K_{\alpha2}$; (3) La $K_{\alpha1}$; (4) La $K_{\alpha2}$; (5) Ce $K_{\alpha1}$; (6) Ce $K_{\alpha2}$; (7) Pr $K_{\alpha1}$; (8) Pr $K_{\alpha2}$; (9) Nd $K_{\alpha1}$; (10) Nd $K_{\alpha2}$; (11) Ba $K_{\beta1}$; (12) Ba $K_{\beta2}$; (13) La $K_{\beta1}$. (14) the original spectrum obtained by the energy dispersive analyzer; (15) approximation of the background radiation; (16) simulation of the K_β components of the Ba and La lines; (17) true spectrum after the subtraction of the K_β components.

After the optimum selection of the K_β components for Ba and La, their lines were subtracted from the original spectrum by the least squares method. The difference represents the true intensity spectrum of the Pr and Nd K_α lines (Fig. 3C). An analogous procedure for the determination of the true spectrum in the region of the Sm lines is shown in Fig.4.

None of the soil reference standards was certified for Pr, Nd, and Sm; only three rock standards were certified for Nd; and one rock standard was certified for Pr and Sm. Therefore, the correlation between the intensities of the spectral lines and the concentrations of the elements cannot be found by conventional methods.

We proposed a generalized calibration curve to be used. The values of the analytical parameter (the ratio between the spectral line intensity and the incoherently scattered exiting radiation intensity) determined by the standard background method as functions of the La, Ce, Pr, Nd, and Sm concentrations for the set of reference standards are shown in Fig. 5. It can be seen that the points for La and Ce (elements sufficiently certified in the reference standards) lie on the generalized curve with good accuracy. All the known points for Pr and Nd also lie on the generalized curve. It can be concluded that the relationship between the analytical parameter and the element concentration will be the same for five lanthanides: La, Ce, Pr, Nd, and Sm. This conclusion is well founded if the closeness of the lines of these elements in the energy spectrum and the remoteness of the exiting radiation are taken into consideration.

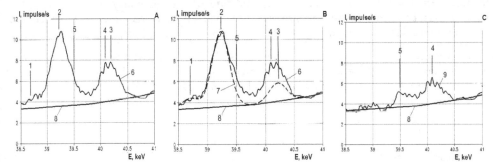

Fig. 4. Spectral segment for the Sm identification (the same sample as in Fig. 3): (A) general spectrum view; (B) detailed interpretation of the spectrum with the simulation of the K_β components of the La and Ce lines; (C) the spectrum after the subtraction of the K_β components. The spectral lines: (1) La $K_{\beta2}$; (2) Ce $K_{\beta1}$; (3) Ce $K_{\beta2}$; (4) Sm $K_{\alpha1}$; (5) Sm K_{u2}. (6) the original spectrum obtained by the energy dispersive analyzer; (7) simulation of the K_β components of the La and Ce lines;(8) approximation of the background radiation; (9) true spectrum after the subtraction of the K_β components.

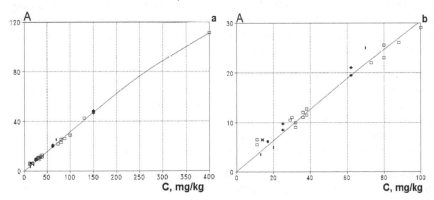

Fig. 5. The analytical parameter (the ratio between the spectral line intensity and the incoherently scattered exiting radiation intensity) as a function of the element concentration: (□) La; (+) Ce; (×) Pr; (ı) Nd; (·) Sm. (a) general view for the set of reference standards; (b) detailed view of the initial part of the curve.

The error of determining Pr, Nd, and Sm cannot be estimated by the conventional method because of the insufficient certification of the reference standards for Pr, Nd, and Sm. The estimation can be based only on indirect considerations. The previously determined values of the K-factors are 1.02 for La and 1.07 for Ce . This implies that, first, the determination the error of these elements almost completely consists of the determination of the error of the spectral line intensity; i.e., the relationship between the analytical parameter and the element concentration (generalized calibration curve) is selected adequately. Second, the values of the K-factors show that an almost ultimate accuracy was reached in the measurements of the line intensities. For the Poisson process of the pulse counting in an energy dispersive analyzer, the absolute error of the intensity measurement is equal to the square root of the intensity. The accuracy of the intensity measurements for Pr, Nd, and Sm will obviously be worse than those for La and Ce, because the intensity of the latter elements is measured after the subtraction of the superimposed lines; the value of the K-factor for them can be estimated as 2.

The metrological parameters were then improved by increasing the time of the spectrum accumulation. The improvement of the parameters was proportional to the square root of the accumulation time. For the samples with low contents of the elements determined, the time of the spectrum accumulation in the ReSPEKT analyzer was increased to 1.5 h.

3. Key results

3.1 Background soils

Podzolization is one of the leading elementary soil forming processes in the forest zone. It is considered to develop due to metal complexing with organic ligands. At present, the following main properties of the podzolic horizon are distinguished: (1) the light gray and whitish color of the horizon (instead of a red or yellowbrown one) due to Fe and Mn removal and the residual accumulation of silica; (2) the acid soil reaction of this horizon and the substantial base unsaturation; and (3) the depletion in nutrients, sesquioxides, and clay particles (Kaurichev, 1989). Fe and Al migrate with the organic complexes from the eluvial to the illuvial horizon. This is also true for K, Na, Ca, and Mg. The podzol formation is assessed according to the degree of the profile differentiation of the iron and aluminum. Special attention is historically paid to Fe, Al, K, Na, and Mg, because the investigation of the soil chemical composition started with measurements of the total content of macroelements.

Later, the development of the cheap express method of X-ray fluorescence analysis permitted obtaining reliable data on the content of heavy metals in podzolic soils. However, the study with this method was somehow misbalanced, as not all heavy metals were involved in investigation in podzolic soils, and an important group of rare earth metals (yttrium and lanthanides) was omitted (Perel'man, 1975). The situation somewhat improved after researchers began to study lanthanides (La and Ce) using X-ray radiometric analysis.

The analysis of the redistribution of specific heavy metals (including the rare earth elements) upon podzolization will facilitate the better understanding of the complex soilformation process and revealing its additional identification criteria.

We studied the taiga soils with eluvial–illuvial profile differentiation in two regions of the European part of Russia, i.e., Arkhangel'sk oblast (the northern taiga) and the Perm region (the southern taiga).

Two soils were investigated in the Pinega district of Arkhangel'sk oblast: an ironilluvial podzol formed on alluvial sandy deposits of a river terrace (further referred to as the *podzol*) under a 40-yearold pine forest that was transformed by numerous fires. A podzolic contact bleached soil with a podzol microprofile (further called the *podzolic soil*) formed on twolayer moraine (loamy sand / sandy heavy loam) under a spruce forest. The E_{pyr} horizon in the podzol contains many charcoal inclusions, and different horizons (BHFn and BF) are identified at a depth of 5–15 cm at different profile walls. The podzol formed on sand is characterized by an acid and weakly acid (in the BC horizon) soil reaction. In the podzolic soil on the two member deposits, the soil reaction varies from acid (to a depth of 16 cm) to neutral (deeper than 30 cm). Some properties of these soils are described in detail in (Goryachkin, Pfeifer, 2005).

The soddypodzolic loamy soil was studied in the town of Chusovoi in the Perm region in the woodland park on the left bank of the Chusovaya River. The soil is acidic to a depth of 85 cm, and deeper the soil reaction is neutral.

Podzol, Pinega. This soil is specified by a low content of heavy metals, which is below their clarkes. Even the illuvial horizons are depleted in Ni, Cu, Ga, Zr, Pb, Fe, Al, Sr, Ba, Y, La, and Ce. The low content of heavy metals is controlled by the light texture of the podzol.

Podzolic soil, Pinega. The content of heavy metals is higher in this soil than in the sandy podzol. The list of metals whose content is below the clarke is much shorter: Ni, Zr, Pb, Sr, Y, and La (Table 2). This is related to the heavy soil texture and, consequently, the lower initial sorting of the moraine substratum.

The soddypodzolic soil in the town of Chusovoi. The content of heavy metals is the highest in this soil. The list of metals whose content is below their clarke values is very short—Zr, Sr, Y, and Ba (Table 2)—which is explained by the heavy texture of the soil.

The average content of these metals in the podzolic soils is below their clarke values. The content of Ce is the highest (29 mg/kg), La ranks second (18 mg/kg), and Y ranks third (16 mg/kg). The higher content of La than Y in the soils is worth noting, although the clarke of Y equal to 40 mg/kg (we consider it to be overestimated) exceeds the clarke of La (34 mg/kg). In this connection, let us point out that the clarke of Y is only 25 mg/kg in the soils of the USA (Ivanov, 1997). The geochemical closeness of Y and La mentioned in publications is proved by the high correlation coefficient between them in the podzolic soils ($r = 0.90$).

We may state that the podzolic soils are depleted in the rareearth metals. In spite of this, the rareearth metals respond readily to the soil podzolization.

As seen in Table 2, the metals differ significantly by their participation in the podzolization process. Let us divide them into three groups according to the illuviation index value. $K_{ill} >$ 5 for the metals participating actively in podzolization; $5 > K_{ill} > 1.5$ for the metals that are moderately active, and $K_{ill} < 1.5$ for the metals virtually not participating in this process.

Let us list the data about the illuviation coefficients of the metals into a summary table (Table 3) in which the metals are divided into three groups according to the degree of their participation in the eluvial–illuvial profile differentiation. This table also manifests the variation coefficients of the illuviation index K_{ill}. They show the variation degree of the eluvial–illuvial distribution of the same metals in different soils. This is evidently related to the different share of reactive metal particles in the different soils.

Horyzon	Depth, cm	Y	La	Ce
Podzol, Pinega				
E_{pyr}	0-2	10	7	10
E	2-5	5	5	6
BHFn	5-13	9	11	12
BF	6-10	10	13	17
BC	13-36	7	7	9
Bce	36-49	8	12	20
D	49-110	11	15	24
Clarke		40	34	43
$C_{BF}:C_{el}$		2.0	2.6	2.8
Podzolic soil, Pinega				
Ele	0-6	14	13	18
ELf	6-16	17	14	19
2EL	16-20	16	21	37
2ELBT	20-30	20	24	44
2BT1	30-65	22	29	53
2BT2	65-90	24	33	54
2BC	90-130	20	31	47
Clarke		40	34	49
$C_{ELBT}:C_{el}$		1.4	1.8	2.4
Soddypodzolic soil, Chusovoi				
AY	3-12	18	15	23
EL	12-27	18	21	30
BEL	27-59	22	21	36
BT1	59-85	36	38	53
BT2	85-104		30	40
C	104-150	23	20	32
Clarke		40	34	43
K_{ill}		2.0	2.5	2.3

Table 2. Content of lanthanides (mg/kg) in the studied soil (podzol, Pinega; podzolic soil, Pinega and soddypodzolic soil, Chusovoi).

Factors influencing the profile redistribution of heavy metals. We studied the influence of the following factors: the pH, the content of clay particles, the organic carbon, and the iron (both total and oxalatesoluble). We calculated the pair correlation coefficients between these soil parameters and the content of lanthanides (using Excel-2007-software). To rule out the impact of the lithogenic factor, the statistical bonds were considered only for the upper horizons. The reliability level was taken for probability of $P = 0.95$ (Table 4).

Metal	Values	Mean	Variation coefficient, %
Highly active metals			
Mn	33; 4.5; 1.3	12.9	135
Fe	4.7; 4.2; 18	9.0	87
Cr	12; 5.9; 1.4	6.4	83
Zn	14; 3,7; 1.3	6.3	106
Moderately active metals			
Cu	4.3; 4.5; 2.1	3.6	37
Ni	4.0; 2.3; 2.8	3.0	29
Ce	2.8; 2.4; 2.3	2.5	11
La	2.6; 1.8; 2.5	2.3	19
Y	2.0; 2.0; 2.0	2.0	0
Pb	2.5; 1.8; 1.5	1.9	27
Al	2.5; 1,4; 1.5	1.8	34
Zr	3.3; 1.2; 1.0	1.8	70
Ga	1.4; 1.0; 2.3	1.6	41
Inert metals			
Sr	1.4; 1.1; 1.0	1.2	17
Ba	1.3; 1.1; 1.0	1.1	14

Table 3. Values of the illuviation coefficient K_{ill} of the metals in the soils of the podzolic group (podzol, Pinega; podzolic soil, Pinega and soddypodzolic soil, Chusovoi).

In the sandy podzol lanthanum is accumulated in the fine silt. In the texturally differentiated podzolic soils the content of lanthnides depends on the pH of the soil solution. The contents La and Ce (which respond to the acidification of soil in the podzolic horizon) depend reliably on the pH value. These metals are cationogenic, with their mobility growing upon the soil's acidification. The content Y is also reliably connected to the amount of clay particles. In soddypodzolic soil, the content of lanthanides Y, La and Ce depends on the soil ferrugination; these metals act as siderophores.

The classic definition of podzolization as a process of iron and aluminum oxides destruction and removal of the decay products should be supplemented by the phenomena of leaching of a number of heavy metals. In addition to Fe and Al, many heavy metals manifest well pronounced eluvial–illuvial redistribution in podzolic soils (Mn, Cr, Zn, Cu, Ni, Ce, La, and Y). The dimensions of these heavy metals' redistribution exceed that of Al. The inactive participation of Al in the redistribution is explained by the insignificant share of its reactive fraction. Although the soils of the podzolic group are depleted in the rare earth metals, the latter readily respond to soil podzolization. In a sandy podzol, the degree of leaching of such heavy metals as Mn, Cr, Zn, Ni, and Zr is markedly higher than in loamy podzolic soil. The leaching of heavy metals from the podzolic horizons has a diagnostic significance, whereas the depletion of metals participating in plant nutrition and biota development is of ecological importance.

Factor	Y	La	Ce
Podzol, Pinega			
pH$_{water}$	-0.08	0.40	0.33
C$_{org}$	0.64	0.04	0.10
Fe$_{tot}$	0.52	0.77	0.59
Clay	0.49	0.42	0.29
Fine silt	0.55	0.94*	0.83
Medium silt	0.52	0.85	0.86
Coarse silt	0.44	0.53	0.55
Fine sand	0.48	0.67	0.75
Medium and coarse sand	-0.55	-0.72	-0.78
Podzolic soil, Pinega			
pH$_{water}$	0.82	0.98*	0.98*
C$_{org}$	-0.56	-0.87	-0.90
Fe$_{tot}$	0.74	0.48	0.45
Clay	0.95*	0.78	0.74
Fine silt	0.64	0.36	0.33
Medium silt	-0.75	-0.84	-0.81
Coarse silt	-0.42	-0.23	-0.16
Fine sand	-0.96	-0.76	-0.74
Medium and coarse sand	-0.91	-0.70	-0.65
Soddypodzolic soil, Chusovoi			
pH$_{water}$	-0.18	-0.36	-0.37
C$_{org}$	-0.43	-0.52	-0.61
Fe$_{tot}$	0.94*	0.89*	0.96*
Clay	0.85	0.86	0.93*
Fine silt	-0.60	-0.40	-0.49
Medium silt	-0.81	-0.64	-0.76
Coarse silt	-0.46	-0.59	-0.61
Fine sand	-0.52	-0.69	-0.68
Medium and coarse sand	-0.56	-0.57	-0.69

*reliable for $P = 95\%$.

Table 4. Pair correlation coefficients between the content of lanthanides and the pH$_{water}$, the content of clay particles, and the organic carbon in the soils' studied (podzol, Pinega; podzolic soil, Pinega and soddypodzolic soil, Chusovoi).

Leaching of heavy metals is most closely related to the destruction of clay particles (in heavy textured podzolic soils in particular); the soil's acidity's influence is less noticeable.

3.2 Soils from natural positive geochemical anomalies

3.2.1 Tundra soils of the Kolyma Lowland

Gley and peat gley cryozems were sampled in tundra of the Kolyma Lowland. Nineteen samples of the fine earth were studied. The soils were described in detail in (Vodyanitskii, Mergelov et al., 2008).

Data on the concentrations of La and Ce in these soils are given in Table 5, and their statistical characteristics are presented in Table 6. The mean element concentrations are high. The accumulation of lanthanides is even more pronounced. Thus, the average concentrations of La and Ce in cryozems reach 46 and 79 mg/kg, which is considerably higher than the clarke values of these elements in the pedosphere (26 and 49 mg/kg) and in the lithosphere (35 and 66 mg/kg, respectively). Hence, the tundra zone of the Kolyma lowland is an example of a positive geochemical anomaly of lanthanides.

Horizon	Depth, cm	La	Ce
Gleyed cryozem, pit T6P2			
Gox	40-50	49	79
Cr(Cg)	50-70	49	75
Gleyed cryozem, pit T8P1			
Gox	3-10	47	74
CR	10-16	48	72
	16-33	46	71
Gox	33-42	51	71
Cr(Cg)	42-52	44	66
Gleyed cryozem, pit T9P3			
CR	0.5-3	43	66
CRg"	3-10	45	66
CR	10-35	42	64
CRg'''	35-54	46	71
Cr(Cg)	54-59	49	71
Gleyed cryozem, pit T10P3			
CRg"	8-18	44	69
	18-35	50	74
Cr(Cg)	58-60	46	70
Peat gleyed cryozem, pit T9P5			
CRg"	12-23	48	69
Cr(Cg)	23-35	46	67
Peat gley soil, pit T9P7			
Gox	10-17	42	59
	17-45	46	71

Table 5. The contents of La and Ce in cryozems of the Kolyma Lowland, mg/kg.

The loss of lanthanides from the soil profile is favored by peat formation and gleyzation. For Ce, the mean concentrations in the cryozems and in the gley horizons of peat gley soils are equal to 71 and 66 mg/kg; and for La - 47 and 45 mg/kg, respectively. However, these differences are statistically unreliable and may be considered a weakly pronounced tendency.

It should be noted that the concentrations of La and Ce in the studied tundra soils are low variable (Table 6): the variation coefficients are equal to 5-6%, which points to the weak differentiation of these elements in the soil profiles.

Element	Mean	Range	Variation coefficient, %
	mg/kg		
Gleyed cryozems			
Lanthanum	47	43-51	5.7
Cerium	71	64-75	5.6
Peat gley soils			
Lanthanum	45	42-48	5.6
Cerium	66	59-71	8.0

Table 6. Statistical characteristics of the La and Ce contents in cryozems of the Kolyma Lowland.

3.2.2 The Kola Peninsula soils

The contents of lanthanides in the soils were determined in northern taiga region: the Khibiny–Lovozero province (Kola Peninsula), where the soils of the background area were studied. On the eastern bank of Lake Umbozero, peatpodzolic soils were sampled (profiles 10 and 11); in the region of a geochemical anomaly due to the closeness to a deposit of loparite ores, samples were taken in three places: a soddy podbur was sampled on the western bank of Lake Lovozero (profile 2); a peaty podzolic soil was sampled on the northern bank of Lake Seidozero (profile 5), and a podzol was sampled on the bank of the Elmoraiok River (profile 9). All the soils were acid (pH$_{water}$ 3.6–5.6) with light sandy and loamy sandy textures. Only the mineral horizons of the soils were analyzed (Table 7).

The background soils and those of the geochemical anomaly formed under the effect of the Lovozero deposit of loparite ores were studied. The content of rare earth metals in loparite is very high (up to 35% in terms of the element oxides). The average chemical composition of loparite with the conventional formula $NaCeTi_2O_6$ (Ivanov, 1997) is as follows (in terms of oxides, %): rare earth metals - 30; Ti - 40; Nb - 12; Na - 8; Sr - 3; Ca - 5; and Ta - 0.8. Among the rare earth elements, Ce is predominant (rel. %): Ce - 49.6; La - 28.4; Pr - 3.4; Nd - 15.5; and Sm - 2.4.

Because of the enrichment of loparite with rare earth metals, their contents exceed the clarke values for the earth's crust by many times: Ce - 133900 : 66 = 2030; La - 76 700 : 35 = 2190; Pr - 9180 : 9 = 1020; Nd - 41850 : 40 = 1050; and Sm - 6480 : 7 = 926. Thus, the loparite containing parent rocks can be significantly enriched with lanthanides, especially the lightest ones (Ce and La). The enrichment of loparite with Pr, Nd, and Sm is twice lower, although it reaches 1000 times. The geochemical anomaly can be heterogeneous because of the different occurrence depths of loparites.

Table 7 shows that the content of Pr in the upper horizons of the soils in the background area is lower than the clarke value by 2 times, and that of Nd by 3 times; the content of Sm is below the detection limit (lower than 1 mg/kg). An eluvial distribution of lanthanides was observed in some of the profiles studied. A similar situation was previously observed in podzolic soils of Sweden and the Kola Peninsula (Nikonov et al., 1999; Tyler, 2004b). Based on the hypothesis about the original layer structure, Sweden soil scientists estimated the losses of lanthanides. A podzol lost 40–50% of the initial contents of Y, La, Ce, Nd, Pr, and Sm over 14000 years (Tyler, 2004a). An analogous calculation shows that 36–47% of the Nd and 0–54% of the Pr were leached from the E horizon of peatpodzolic soils in the Khibiny–Lovozero province; 53% of the Nd, 60% of the Pr, and 50% of the Sm were leached from the E horizon of a podzol. The instability of lanthanides in the acid soils of the taiga zone was confirmed.

Horizon	Depth, cm	Pr	$Pr_s : Pr_c$	Nd	$Nd_s : Nd_c$	Sm	$Sm_s : Sm_c$	Pr : Nd	Pr :Sm
Background Profile 10. Umbozero bank. Peatpodzolic soil									
E	12-18	4	0.53	8	0.42	-	-	0.50	-
BT	18-28	2	0.26	7	0.37	-	-	0.29	-
C	28-47	4	0.53	15	0.79	-	-	0.27	-
Profile 11. Umbozero bank. Peatpodzolic soil									
E	8-10	3	0.42	7	0.37	-	-	0.43	-
BT	10-29	3	0.42	6	0.32	-	-	0.50	-
C	29-49	3	0.42	11	0.58	-	-	0.27	-
Geochemical anomaly Profile 2. Lovozero bank. Soddy podbur									
AY	5-12	14	1.8	44	2.3	11	2.4	0.32	1.27
BF	12-40	13	1.7	57	3.0	10	2.2	0.23	1.30
BC	40-50	12	1.6	59	3.1	10	2.2	0.20	1.20
Profile 5. Seidozero bank. Peatpodzolic soil									
E	13-24	21	2.7	104	5.5	18	4.0	0.20	1.17
BT	24-40	30	3.9	142	7.5	23	5.1	0.21	1.30
C	40-62	46	6.0	198	10.4	35	7.8	0.23	1.31
Profile 9. Elmoraiok River bank. Podzol									
E	5-15	22	2.9	106	5.6	16	3.5	0.21	1.37
BT	15-30	36	4.7	156	8.2	26	5.8	0.23	1.38
C	30-52	55	7.2	224	11.8	32	7.1	0.24	1.72
Clarke (c)		9		40		7.0		0.22	1.29
Clarke (s)		7.6		19		4.5		0.40	1.69

Table 7. Pr, Nd, and Sm in the mineral horizons of soils in the Khibiny–Lovozero province, mg/kg.

In the Lovozero geochemical anomaly, the contents of Pr, Nd, and Sm in the mineral horizons exceed the clarke values. In the soddy podbur on the Lovozero bank, their contents are higher than the clarkes by 1.6–1.8 times for Pr, 2.2–2.4 times for Sm, and by 2.3–3.1 times for Nd. The concentrations of lanthanides in the peatpodzolic soil on the bank of Lake Seidozero and in the podzol on the bank of the Elmoraiok River are even higher. Here, the clarke values are exceed by 2.7–7.2 times for Pr, by 4–7.8 times for Sm, and even by 5.5–11.8 times for Nd. Thus, the soils are strongly enriched with these lanthanides, especially Nd.

The statistical relationships with a wide range of lanthanides studied previously were considered. In the background area, Y was related to La and Ce, and La was related to Ce with high reliability (r = 0.83–0.96). The content of Pr, on the contrary, was not related to any lanthanide. The peculiar behavior of Pr in the soil can be related to its increased biophilicity.

Let us discuss the reason for the difference in the distributions of Pr and Nd, although their contents in loparite, the main rare earth mineral of the province, are similar with consideration for their clarkes. The difference in the biophilic properties of the lanthanides is a probable reason. No reliable data are available on the biophilic properties of lanthanides. The coefficients of the biological absorption of lanthanides are absent in the revue of Perel'man (1975). This group of elements was called biotrophic (biophilic). Phosphorus is a known biophil with a high coefficient of biological absorption (~100) (Perel'man, 1975); therefore, it can be taken that lanthanides are also biophils for some plants. This is also true for mosses and peatforming plants in the Khibiny–Lovozero province. According to Dyatlova et al. (1988), the content of Sm in the moss and lichen ash is 2–40 mg/kg (its clarke is only 7 mg/kg), and the content of Nd is 8– 150 mg/kg (its clarke is 19 mg/kg). This confirms the biophilic character of lanthanides.

The degrees of biophilicity of lanthanides can be different. The comparison of the Pr and Nd clarkes in the soil and the earth's crust indicates the different degrees of their global persistence in soils. The index of accumulation/dispersion in the soil, IA = Clarke (soil) : Clarke (earth crust), is 19 : 40 = 0.47 for Nd and 7.6 : 9 = 0.84 for Pr. Our data confirm the thesis about the higher biophilicity of Pr compared to Nd. The analysis of the ash of lowash peats from the surface of peatpodzolic soils showed a difference in the biological absorption coefficients A_x (against the earth's crust clarke) of these two lanthanides: A_{Pr} = 1.1 and A_{Nd} = 0.5 on the average for the background area; A_{Pr} = 3.3 and A_{Nd} = 2.7 for the anomaly area. There is no reliable correlation between the contents of Pr and Nd: r = 0.47.

In the profile of the peatpodzolic soils in the background area and the soddy podbur in the anomaly, the distribution of Pr is uniform, and that of Nd is eluvial, which explains the abovenoted absence of a statistical correlation between the contents of Pr and other lanthanides in the mineral layers. This can be due to the relative (biogenic?) accumulation of Pr in the upper layer of soils with relatively homogeneous textures: sandy soils in the background area and loamy sandy soils in the anomaly.

3.3 Contaminated soils

3.3.1 Urbanozems and soddy-podzolic soils of Perm

Urban soils (urbanozems) and soddy-podzolic soils from the city of Perm were studied. Samples from surface (0–20 cm) layers of soddy-podzolic soils were taken in parks in the central part of the city. Samples of urbanozems from the same depth were also taken on the lawns along roadways polluted by heavy metals in the central part of the city. To compare distribution patterns of the studied elements with those of well-known air pollutants, the contents of lead and nickel were determined in these soils by the routine X-ray fluorescence analysis. In total, twenty soil samples were studied.

In addition to La and Ce, the concentrations of Ba, Ni and Pb in the surface layers of these soils were determined, and their distribution patterns were compared. They differed for the two groups of metals (Table 8). The concentrations of Ni and Pb in these soils vary considerably, which is typical of urban soils, including soils of Perm (Eremchenko, Moskvina, 2005). The variation coefficients are 53% for Ni and 126% for Pb. This is explained by the fact that the samples were taken not only in parks with soddy-podzolic soils but also from urbanozems near roads with intensive traffic. Variations in the contents of Ba, La, and Ce are considerably lower (V = 21–33%). Thus, the portion of metals of technogenic origin for these three elements is much lower than that for Ni and Pb.

Element	Mean	Range	Variation coefficient, %
	mg/kg		
Nickel	145	31–268	53
Lead	113	24–630	126
Barium	382	301–627	21
Lanthanum	19	12–31	28
Cerium	26	18–44	33

Table 8. Statistical characteristics of the Ni, Pb, Ba, La and Ce contents in urbanozems and soddy-podzolic soils in Perm.

Despite the presence of technogenic elements, average concentrations of Ba, La, and Ce are low and remain below the clarke values, which attests to the removal of lanthanides and Ba from the soil profiles. This regularity is especially pronounced in comparison with the soils of the Kolyma lowland. In the automorphic soils of Perm, the mean Ba and lanthanides' contents are by 1.6 and 2.4–2.7 times lower than those in the tundra soils of the Kolyma Lowland. Thus, active weathering and the removal of Ba, La, and Ce under conditions of humid climate are typical of these automorphic soils.

3.3.2 Alluvial soils of Perm

Meadow-bog alluvial soils were studied on floodplains of small rivers and the Kama River in Perm. Overall, 16 samples of the fine earth and nodules and rhyzoconcretions from these soils were examined. To judge about the accumulation or depletion of Ba, La, and Ce in the nodules and rhyzoconcretions, the coefficient of element accumulation (C_{ac}) in them was calculated: $C_{ac} = C_{concr} : C_{fine}$ earth, where C_{concr} and C_{fine} earth are the contents of particular elements in the concretions and fine earth, respectively. The studied soils were described in (Vodaynitskii, Vasil'ev et al., 2007).

We studied fine earth and iron–manganic concretions (nodules and rhyzoconcretions) (Table 9). The content of Ba in the fine earth of soils varies from 406 to 527 mg/kg, while its content in the concretions varies greatly (from 588 to 2848 mg/kg). The coefficient of Ba accumulation in the concretions also varies considerably: from 1.2 to 6.0 barium fixation in them. Light-colored formations with a size of about 20 µm were identified on their internal surface. The microanalyzer showed that these formations of composed of barite ($BaSO_4$). Neoformations of barite were earlier determined by Bronnikova and Targulian (2005) on the surface of cutans in podzolic soils. Barium is not typical of the soils of forest landscapes, so its accumulation in the cutans is related to the soil pollution. In the nodules from the alluvial soils, Ba is mainly accumulated as barite particles precipitated on the active matric of the nodules.

In the gleyed alluvial agrozem on the floodplain of the Mulyanka River polluted by wastewater, Ba redistribution in the soil profile takes place. In the fine earth, the Ba content

Concepts and Principles of Geochemistry

is relatively stable (421–474 mg/kg). In the nodules, the concentration of Ba varies considerably and reaches its maximum (2840 mg/kg) in the C4g,t horizon at a depth of 107–137 cm. In this coarse-textured alluvial soil, Ba migrates easily to a considerable depth and is concentrated in the nodules.

Horizon; depth, cm	Substrate	Ba	C_{ac} (Ba)	La	C_{ac} (La)	Ce	C_{ac} (Ce)	Ce/ La
Typical humus-gley soil on the Obva River floodplain, pit 51								
C2g, 37-75	Fine earth	527		37		58		1.6
	Rhyzoconcretions	623	1.2	25	0.7	38	0.6	1.5
G~~, 75-90	Fine earth	523		41		60		1.5
	Rhyzoconcretions	673	1.3	14	0.3	16	0.3	1.1
Stratified typical alluvial soil on the Obva River floodplain, pit 53								
C2~~, 20-27	Fine earth	410		28		41		1.5
C6~~, 71-78	Fine earth	406		26		38		1.5
Mineralized humus-gley soil on the Kama River floodplain, pit 41								
G~~, 31-55	Fine earth	452		38		57		1.5
	Nodules	715	1.6	56	1.5	191	3.3	3.4
Gleyed agrozem on the Mulyanka River floodplain, pit 33								
C2~~, 49-75	Fine earth	430		31		47		1.5
	Nodules	195 7	4.5	104	3.3	324	6.9	3.1
C3~~, 75-107	Fine earth	421		31		46		1.5
	Nodules	2120	5.0	108	3.5	302	6.6	2.8
C4g,t~~, 107-137	Fine earth	474		34		45		1.3
	Nodules	2840	6.0	100	2.9	243	5.4	2.4
C5g~~, >137	Fine earth	441		30		48		1.6
	Nodules	588	1.3	86	2.9	150	3.1	1.7
Average			3.0		2.2		3.7	

Table 9. The contents of Ba, La, and Ce (mg/kg) in the fine earth and nodules of alluvial soils in the Cis-Ural region, coefficients of metal accumulation in the nodules (C_{ac}), and the Ce-to-La ratios (Savichev, Vodyanitskii, 2009).

The La content in the fine earth varies from 28 to 41 mg/kg. In the nodules, it varies from 14 to 108 mg/kg, and the coefficient of La accumulation in the nodules varies from 0.3 to 3.5. In the gleyed alluvial agrozem studied on the floodplain of the Mulyanka River polluted by wastewaters, including those from the petroleum refinery, the content of La in the fine earth remains practically stable (30–34 mg/kg). Its content in the nodules changes from 86 mg/kg in the deepest C5g horizon to 108 mg/kg in the C3 horizon at a depth of 75–107 cm. The technogenic lanthanum migrates to a considerable depth in this coarse-textured soil.

The Ce content in fine earth varies from 38 to 60 mg/kg. In the nodules, it varies from 16 to 324 mg/kg. The corresponding coefficients of Ce accumulation in the nodules vary from 0.3 to 6.9. In the gleyed alluvial agrozem on the floodplain of the Mulyanka River, the Ce content in the fine earth remains almost stable throughout the soil profile (45–48 mg/kg). In the nodules, it changes from 150 mg/kg in the deepest C5g horizon to 324 mg/kg in the C2 horizon at a depth of 49–75 cm. Being most active among lanthanides, the technogenic Ce migrates down the soil profile to a relatively shallow depth and is fixed in the nodules. This is also confirmed by other data. Thus, in the tropical laterites of Cameron, lanthanides are removed from the top iron-enriched horizons and are accumulated in the deeper layers. Cerium is deposited just below the eluvial horizon, and other lanthanides are accumulated in the deeper horizons (Braun, Viers et al., 2005). Similar differentiation of lanthanides is seen in the soil profiles on granodiorites in the New Southern Wales (Australia).

It should be noted that the contents of La and Ce in the studied soils of Perm region remain below the clarke values for the pedosphere despite the soil pollution (Taunton, Welch et al., 2000).

Different mechanisms of the formation of rhyzoconcretions and iron nodules are responsible for the difference in the accumulation coefficients of the three elements. Rhyzoconcretions formed with participation of organic root exudates on the floodplain of the unpolluted Obva River are characterized by the weak Ba accumulation (C_{ac} = 1.2–1.3) and the depletion of La and Ce (C_{ac} = 0.3–0.7). Organic ligands in the rhyzoconcretions are mainly spent for iron fixation. This is reflected in the wide Fe-to-Mn ratio in them (25–100). Iron–manganic nodules are formed under conditions of alternating redox regime and are enriched in all the three elements. They are characterized by a more even accumulation of elements and a lower Fe-to-Mn ratio (1.4–12).

The degree of pollution of the river is also important. In the floodplain soil of the strongly polluted Mulyanka River, the accumulation coefficient of Ba in concretions varies from 1.3 to 6.0, the accumulation coefficient of La reaches 2.9–3.5, and the accumulation coefficient of Ce is as high as 3.1–6.9. Polluted waters of the Mulyanka River enter the deep Kama River and are diluted. As a result, nodules in the soils on the Kama River floodplain have lower concentrations of the studied elements; the coefficients of their accumulation are equal to 1.6 for Ba, 1.5 for La, and 3.3 for Ce.

According to the average coefficients of element accumulation in the concretions, the studied elements form the following sequence: Ce(3.7) > Ba(3.0) > La(2.2). The degree of an element accumulation in the concretion depends on its sensitivity to changes in the redox regime, sorption capacity, and the capacity of the element to form stable complexes with organic ligands. The most pronounced accumulation of Ce in the concretions is explained by its sensitivity for changes in the redox regime. Intermediate position of barium is explained by the precipitation of barite crystals on the active matrix of the concretions; Ba is a manganophilic element. A relatively low accumulation of La in the concretions is explained by its physicochemical inactivity.

Let us compare the Ce-to-La ratios in the fine earth and concretions. In the fine earth, this ratio is practically constant and averages to 1.5, which is close to the clarke value (1.9). In the concretions, it varies from 1.1 to 3.4 and averages to 2.3. A wider Ce-to-La ratio in the concretions signifies that Ce accumulation in them is more active as compared with La.

3.3.3 The soils contaminated with the emissions from the Noril'sk metallurgical enterprise

Noril'sk is the center of the industrial region in the south of the Taimyr Peninsula. Three metallurgical plants are located in the immediate vicinity of the town producing a high technogenic load and a high level of soil contamination. In the town, the lawn soils are formed by mixing metallurgical and coal slags with soil or peat. Lawns are situated above the heating mains laid at the surface. The gas and dust emissions of the metallurgical enterprise affect the soils outside the town.

The soils were sampled in June of 2004. We analyzed the contaminated soils of Noril'sk and its suburbs located at a different distance from the town to the northeast. Mixed samples of the surface soil horizon (0–5 cm) were taken, as well as samples from the genetic horizons of the gley cryozem profile formed on heavy and medium marine loams. The soils develop under the conditions of permafrost occurring close to the surface, which results in the weak evaporation of moisture and the development of gley.

The investigated region is subdivided into two zones according to their contamination, i.e., (1) the urban territory, where the soil contamination is controlled by the slags to a great extent; (2) the suburb area at a distance of 4–15 km, where the soil contamination is controlled by aerosols. The soils in Noril'sk are maximally contaminated. The clarkes of the Cu, Ni, Cr, Zn, Fe, and S are exceeded by 287, 78, 4.7, 3.8, 4.1, and 3.5 times, respectively. In the suburbs, the main pollutants remain the same; however, the pollution's degree is lower; the clarke excess is equal to 65 for Cu, 35 for Ni, and 2.4 for Fe. The town and its suburbs form a technogenic copper–nickel anomaly. The contamination decreases unevenly. As compared to the town, the nickel's contamination decrease is less pronounced in the suburbs than the copper contamination. The slags in the town apparently contain more copper than nickel.

The average content of lanthanides is far below their clarke values in the suburban soils: clarkes of concentration for La and Ce being equal to 0.4. The territory in the town's vicinity represents a negative technogenic geochemical anomaly according the content of these metals (Table 10).

Distance from Noril'sk, km	Horizon	Depth, cm	Y	La	Ce
Noril'sk city					
0	A	0-10	34-23	21-15	30-19
Noril'sk suburb					
4	B	3-70	24	17	28
	G	70-90	33	19	29
6	Mixed	0-10	18	9	13
8	Mixed	0-10	30	14	19
9	Mixed	0-10	19	13	17
10	A	2-4	16	10	15
	B1	4-45	34	19	34
12	Mixed	0-10	18	15	22
14	Mixed	0-10	25	13	20
15	Mixed	0-10	16	12	19
Clarke			40	34	49

Table 10. Content of lanthanides (mg/kg) in the soils of the Noril'sk technogeochemical anomaly.

We may judge about the bearing phases of the pollutants from Table 11, which manifests the correlation coefficients between the content of the heavy metals and the iron and sulfur. Chromium and zinc operate as siderophilic elements in the urban soils, which is explained by the soil's contamination with slags containing magnetite. The high magnetic susceptibility of the soils (840×10^{-8} m^3/kg) points to the presence of magnetite. The content of the principal pollutants (nickel and copper) correlates with that of sulfur proving the sulfidic nature of the technogenic Ni and Cu. It is no surprise because the copper–nickel deposits in Noril'sk and Talnakh are of the sulfide type (Ivanov, 1997).

Correlation	Y	La	Ce
Noril'sk city			
Fe - Ln	-0.27	-0.06	-0.43
S - Ln	-0.63	-0.22	-0.73
Noril'sk suburb			
Fe - Ln	0.76*	0.82*	0.87*
S - Ln	-0.67*	-0.66*	-0.65*

*Reliable at $P = 0.95$.

Table 11. Correlation coefficients between the content of sulfur and iron and the content of lanthanides (Ln) in the soils of Noril'sk ($n = 14$) and its suburbs ($n = 10$)

The situation is more difficult for the suburban soils. Two opposite processes are effective there: the siderophilic properties are intensified for some elements, and the chalcophilic properties for others. Lanthanides manifest siderophilic properties. This phenomenon is controlled by the increasing influence of the background conditions, under which the content of lanthanides depends on the iron (hydr)oxides. The comparison of the correlation coefficients between lanthanides and Fe in the town (where they are negative) and in the suburbs (where they are positive and high) leads us to the conclusion that the lanthanides enter the soil from the slags rather than from natural bodies in the town. We may speak about the technogenic origin of the lanthanides, although this conclusion cannot be drawn from other more rough indices.

4. Conclusion

1. X-ray fluorescence method is the simplest and cost-effective method for studying heavy metals in soils. However, under normal working conditions, niobium with $Z = 41$ is the last element that may be identified by this method. The use of X-ray radiometric method makes it possible to determine more elements. In this method, a sample is excited not by emission of the X-ray tube, but by the radioisotope source with a great radiant energy. We have developed the methods for determining Ba, La, and Ce contents with the use of [241]Am isotope source. The new method has made it possible to obtain data on geochemistry of Ba, La and Ce in soils of humid landscapes.

2. We developed a procedure for the identification of Pr, Nd and Sm using a [241]Am isotope source. The procedure is based on the exclusion of the disturbing effect of Ba and La on the lines of Pr and Nd, as well as the effect of La and Ce on the lines of Sm. On the basis of the new method, data were obtained on the geochemistry of three lanthanides in the northern taiga soils.

3. The classic definition of podzolization as a process of iron and aluminum oxides destruction and removal of the decay products should be supplemented by the phenomena of leaching of a number of heavy metals. In addition to Fe and Al, many heavy metals manifest well pronounced eluvial–illuvial redistribution in podzolic soils (Mn, Cr, Zn, Cu, Ni, Ce, La and Y). The dimensions of these heavy metals' redistribution exceed that of Al. The inactive participation of Al in the redistribution is explained by the insignificant share of its reactive fraction. Although the soils of the podzolic group are depleted in the rare earth metals, the latter readily respond to soil podzolization. In a sandy podzol, the degree of leaching of such heavy metals as Mn, Cr, Zn, Ni and Zr is markedly higher than in loamy podzolic soil. The leaching of heavy metals from the podzolic horizons has a diagnostic significance, whereas the depletion of metals participating in plant nutrition and biota development is of ecological importance.

    Leaching of heavy metals is most closely related to the destruction of clay particles (in heavytextured podzolic soils in particular); the soil's acidity's influence is less noticeable.

    The heavy alkalineearth metals (Sr and Ba) do not participate in podzolization.

4. In the cryozems of the Kolyma Lowland, the high content of La and Ce has been determined. Weathering processes in these cold soils are retarded, which is confirmed by the low role of iron oxidogenesis. This results in preservation of lanthanides and barium in the cryozems.

5. In the Khibiny–Lovozero province of the Kola Peninsula, the area is divided into two parts. In the soils near Lake Umbozero, the contents of Pr and Nd are lower than their clarkes, and the content of Sm is below the detection limit. In the background area, the lanthanides are strongly leached from the podzolic soils with Nd being leached more strongly than Pr.

    In the region of the geochemical anomaly (near Lake Lovozero), the contents of Pr, Nd, and Sm are significantly higher than their clarkes due to the effect of the adjacent deposit of loparites. In the background area, Y is related to La and Ce, and La is related to Ce with high reliability, while the content of Pr is not related to any lanthanide. This is explained by the uniform profile distribution of Nd, in distinction from the eluvial distributions of Y, La and Ce.

    Positive rare earth anomalies can be expected in soils located not far from the deposits of apatitenephelines, loparites and phosphorites and in the soils developed on alkaline granites and carbonate weathering crusts.

6. In the soddy-podzolic soils and urbanozems of Perm, the variability in the contents of Ba, La, and Ce is considerably lower than that of Ni and Pb. The first three elements are characterized by a smaller portion of technogenic elements as compared with Ni and Pb. The mean contents of Ba, La, and Ce are low despite the possible contribution of technogenic sources. Soil weathering accompanied by the removal of these elements under conditions of humid climate, which is responsible for their low content in the soil profiles.

7. In the alluvial soils of Perm, Ba is fixed in rhyzoconcretions as barite ($BaSO_4$) of, probably, technogenic nature. In the gleyed alluvial agrozem on the Mulyanka River

floodplain, Ba migrates intensively in the soil profile and is fixed in iron–manganic nodules. The contents of La and Ce in the fine earth of polluted soils remain practically stable, and lanthanides are accumulated in the concretions: the accumulation coefficient reaches 3.5 for La and 6.9 for Ce.

8. The soil cover in the area influenced by the Noril'sk metallurgical enterprise may be subdivided in terms of the soil contamination into two territories: Noril'sk proper and its suburb at a distance of 4–15 km. The urban soils are maximally polluted with their clarke excesses being equal to 287, 78, 4.7, 4.1, and 3.5 for copper, nickel, chromium, iron, and sulfur, respectively. In the Noril'sk suburbs, the principal pollutants are the same; however, the clarke excess is lower: 65 for Cu, 35 for Ni, and 2.4 for Fe. The urban and suburban territory represents a technogenic copper–nickel–chromium anomaly. By the content of a number of superheavy metals (Ba, La and Ce), the territory near the town forms a negative geochemical anomaly. The situation is different in the suburban soils. The rare earth elements (Y, La and Ce) have pronounce siderophilic properties due to the natural factors' influence.

5. References

Arnautov N.S. (ed.) (1987). Standard Samples of the Chemical Composition of Natural Mineral Substances. Methodological Guidelines, Novosibirsk. [in Russian].

Berenshtein L.E., Masolovich N.S., Sochevanov V.G. & Ostroumov G.V. (1979). Metrological Basis of Quality Control of Analytical Works, In: *Methodological Basis of Studying the Chemical Composition of Rocks, Ores and Minerals*, pp. 23–118, Nedra, Moscow. [in Russian].

Bowen H.J.M. (1979). *Environmental Chemistry of Elements*, Academic, New York.

Braun J.J., Viers J. & Dupre B. (2005). Solid/Liquid REE Fractionation in the Lateritic System of Goyoum, East Cameroon: The Implication For the Present Dynamics of Soil Covers of the Humid Tropical Regions. *Geochim. Cosmochim. Acta*, Vol. 62, pp. 273–299.

Bronnikova M.A. & Targul'yan V.O. (2005). *The Cutan Assemblage in Texture-Differentiated Soils*, Akademkniga, Moscow. [in Russian].

Dyatlova N.M., Temkina V.Ya. & K. I. Popov K.I. (1988). *Complexones and Complexonates of Metals*, Khimiya, Moscow. [in Russian].

Eremchenko O.Z. & Moskvina N.V. (2005). The Properties of Soils and Technogenic Surface Formations in the Multistory Districts of Perm City. *Pochvovedenie*, No. 7, pp. 782–789. [*Eur. Soil Sci.* 38 (7) (2005)], ISSN 1064-2293.

Evans C.H. (1990). *Biochemistry of the Lanthanides*, Plenum, New York.

Goryachkin S.V. & Pfeifer E.M. (ed.) (2005). *Soils and Perennial Underground Ice of Glaciated and Karst Landscapes in Northern European Russia*, Inst. Geogr., Moscow.

Greenwood N.N. & Earnshaw A. (1997). *Chemistry of Elements*, 2nd ed., Elsevier.

Inisheva L.I & Ezupenok E.E. (2007). Contents of Chemical Elements in Highmoor Peats, In: *Current Problems of Soil Pollution: II International Scientific Conference, Moscow, Russia* (Moscow, 2007), Vol. 2, pp. 63–67. [in Russian].

Ivanov I.N. & Burmistenko Yu.N. (1986). *Neutron-activation Analysis and the Use of Short Living Radionuclide*, Energoizdat, Moscow. [in Russian].

Ivanov V.V. (1997). *Ecological Geochemistry of Elements*, Ekologiya, Moscow. [in Russian], ISBN 5-247-03178-4.

Kabata-Pendias A. & Pendias H. (1985). *Trace Elements in Soils and Plants*, CRC, Boca Raton.

Kashulina G.M., Chekushin V.A. & Bogatyrev I.V. (2007). Physical Degradation and Chemical Contamination of Soils in Northwestern Europe, In: *Current Problems of Soil Pollution: II International Scientific Conference, Moscow, Russia* (Moscow, 2007), Vol. 2, pp. 74– 78. [in Russian].

Kaurichev I.S. (ed.) (1989). *Soil Science*, Agropromizdat, Moscow. [in Russian].

Khitrov V.G. (1985). The Results of Attestation of the System of Standards of the Chemical Composition of Magmatic Rocks. *Izv. Akad. Nauk SSSR, Ser. Geol.*, No. 11, pp. 37–52.

Nikonov V.V., Lukina N.V. & Frontas'eva M.V. (1999). Trace Elements in Al–Fe–Humus Podzolic Soils Subjected to Aerial Pollution from the Copper–Nickel Production Industry in Conditions of Varying Lithogenic Background. *Pochvovedenie*, No. 3, pp. 370–382. [*Eur. Soil Sci.* 32 (3), 338–349 (1999)], ISSN 1064-2293.

Perel'man A.I. (1975). *Geochemistry of Landscape*, Vysshaya shkola, Moscow. [in Russian].

Perelomov L.V. (2007). Interactions of Rare Earth Elements with Biotic and Abiotic Soil Components. *Agrokhimiya*, No. 11, pp. 85–96, ISSN 0002-1881.

Savichev A.T. & Fogel'son M.S. (1987). X-ray Fluorescent Analysis of Silicate Rocks on a Spectrometer. *Izv. Akad. Nauk SSSR, Ser. Geol.*, No. 8, pp. 103–108.

Savichev A.T. & Sorokin S.E. (2000). X-Ray Fluorescence Analysis of Microelements and Heavy Metals in Soils. *Agrokhimiya*, No. 12, pp. 71–74, ISSN 0002-1881.

Savichev A.T. & Vodyanitskii Yu.N. (2009). Determination of Barium, Lanthanum, and Cerium Contents in Soils by the X-Ray Radiometric Method. *Eur. Soil Sci.*, Vol. 42, No 13, pp. 1461–1469, ISSN 1064-2293.

Taunton A.E., Welch S.A. & Banfield J.F. (2000). Microbial Control on Phosphate and Lanthanide Distributions during Granite Weathering and Soil Formation. *Chem. Geol.*, Vol. 169, pp. 371–382.

Tyler G. (2004a). Rare Earth Elements in Soil and Plant Systems: A Review. *Plant Soil*, Vol. 267, pp. 191–206, ISSN 0032-079X.

Tyler G. (2004b). Vertical Distribution of Major, Minor, and Rare Elements in Haplic Podzol. *Geoderma*, Vol. 119, pp. 277–290, ISSN 0016-7061.

Vodyanitskii Yu.N., Mergelov N.S. & Goryachkin S.V. (2008). Diagnostics of Gleyzation upon a Low Content of Iron Oxides (Using the Example of Tundra Soils in the Kolyma Lowland). *Pochvovedenie*, No. 3, pp. 261–279. [*Eur. Soil Sci.* 41 (3), 231–248], ISSN 1064-2293.

Vodyanitskii Yu.N., Vasil'ev A.A. & Kozheva A.V. (2007). Influence of Iron-Containing Pigments on the Color of Soils on Alluvium of the Middle Kama Plain. *Pochvovedenie*, No. 3, pp. 318–330. [*Eur. Soil Sci.* 40 (3), 289–301], ISSN 1064-2293.

Wu Z.M. & Guo B.S. (1995). *Bioinorganic Chemistry of Rare Earth Elements*, J. Z. Ni (ed.), Science Press, Beijing, pp. 13–55.

Zhu W.F., Xu S. & Zhang H. (1995). Biological Effect of Rare Earth Elements in Rare Earth Mineral Zone in the South of China. *Chin Sci. Bull.*, pp. 914–916.

6

Application of Nondestructive X-Ray Fluorescence Method (XRF) in Soils, Friable and Marine Sediments and Ecological Materials

Tatyana Gunicheva
A. P. Vinogradov Institute of Geochemistry
Russia

1. Introduction

X-ray fluorescence (XRF) analysis is accepted as the most suitable physical method for the exploration of the elemental composition of rocks and minerals. This is due to fusing the sample with appropriate flux. The desired result is achieved because rocks and minerals are oxidic systems relative to major components. Soils, friable and marine sediments, silt and ecological materials differ from the above in the presence of an organic constituent (Corg), the weight fraction of which may vary considerably. The bioorganic diversity of Corg is the main source of errors, arising from sampling and analyzing procedure (Bock, 1972). Therefore, the possibility to analyze samples without having them destructed (nondestructive), preserving the study material after the results have been obtained, remains the unique advantage of XRF. This chapter reports information on nondestructive XRF procedures to determine the contents of rock-forming and some minor elements in powder of the materials listed. A satisfactory quality of XRF results, their validity and prospective viability for multi-purpose interpretations and environment monitoring have been discussed.

2. Nondestructive X-ray fluorescence (XRF) analysis of soils, friable and marine sediments

Their organic constituents consist of a mixture of plant and animal products, decomposed to different extents, and compounds which are chemically and biologically synthesized in soil. The resultant products of these processes are humic matter, low and high molecular weight organic acids, carbohydrates, proteins, peptides, amino acids, lipids, waxes, polycyclic aromatic hydrocarbons and lignin fragments. In addition, the secretions of root systems, consisting of a wide range of simple organic acids, are also present in the soils. The humic matter has the structure of a twisted polymer chain and consists of a relatively large number of functional groups (CO_2, OH, C=C, COOH, SH, CO_2H). Owing to a specific combination of various groups (particularly OH and SH), the humic matter is capable of producing complex compounds with some cations (Kabata-Pendias & Pendias 1986; Bolt & Bruggenwert 1976; Greenland & Hayes 1978; Lindsay 1979). The composition and properties of the organic constituents of soil depend on climatic conditions, the type of soil and agrotechnical techniques. Their interactions with soil metals may be described with the help of such phenomena as ion-exchange reaction, surface sorption, chelate formation, coagulation and peptization. Biochemical complexity of the organic constituent of the materials considered radically alters melting with the flux. The formation of metal carbides, conversion of organic carbon into its modifications, the combustion point of which is very high (is not below 1600°C) and other phenomena arising in this situation inhibit homogenization. Therefore, the formal utilization of XRF analysis of rocks for soils and sediments seems to be possible only when their Corg. content does not exceed 1.5-2% (Kabata-Pendias & Pendias 1986). In all other instances fusing leads to the isolation of carbon at the glass disc surface. Concerning the materials ashed, because of the large number of non-investigated effects taking place under the recommended ashing at temperature 525 ±25° C conversion from the ashed to the initial system is so uncertain that its analysis becomes meaningless.

This part of paper reports information on the nondestructive XRF determination of Na, Mg, Al, Si, P, K, Ca, Ti, Mn, Fe, S, Ba, Sr and Zr in secondary natural matters. Except for drying at 105° C and pressing, it does not require any preliminary treatment of the sample. The necessity for additional drying is connected with the fact that the certified estimates of composition for standard materials used for calibration are given for those dried and sterilized at 105°C.

2.1 Radiator preparation

Tablets from the powder samples were pressed on a boric acid backing under constant pressure. The amount of material required to produce a specimen for XRF analysis is different because of the varying organic content. Thus, if for soil and loam 6 g are sufficient, then for deposits the amount should not be less than 8 g. For humus and forest litter, its total trace element content does not exceed 5-6%, so that an 'infinitely thick' layer is ensured by about 12-14 g.

2.2 XRF equipment

The intensities of analytical lines were measured with a CPM-25 x -ray spectrometer with 16 fixed channels. The rhodium target x-ray tube was operated at 40 kV. The scattered Rh Kα - line intensity measured with the 16th channel was used as an internal standard for some elements. It should be kept in mind that the wavelength of this line is the shortest among those measured.

2.3 Standard set for calibration

For calibration and assessment of the accuracy of analysis we used sets of Russia national certified standards of soils, marine sediments and friable deposits (Arnautov 1987). With the help of Chinese reference standards of soils and river sediments the possibility of the joint use of the Russia and Chinese national collections was also estimated.

In Table 1 for standard materials of various types of soil, sediment and friable deposits, the results obtained by the proposed XRF procedure (XRF) are compared with the certified values. For Al and Si, the XRF values were calculated using the set of calibration standards restricted to standards of the same type. The agreement between the XRF and certified values is satisfactory (Ostroumov 1979). With the exception of Si and partly for Al, for all elements the differences among them are not significant and do not exceed the permissible standard deviations for all types of the above materials. The discrepancies observed for Al and Si are not surprising. In our opinion, the reason is that these elements are major and present in distinct mineral phases. However, our aim was to show the accuracy of the analytical results which the proposed method will provide for samples prepared in the required way. If the analytical data user is to obtain such results, real-world samples must be treated in the same manner, otherwise the quality of the final results will be significantly worse than the above.

Component	SP-1 black earth soil		SP-2 podzol soil		SKR red earth soil		SSK grey earth carbonate soil	
	Certified	XRF	Certified	XRF	Certified	XRF	Certified	XRF
Na_2O	0.80±0.03	0.81±0.02	1.15±0.05	1.09±0.02	0.15±0.03	<0.2	1.64±0.05	1.76±0.03
MgO	1.02±0.03	1.10±0.02	0.77±0.01	0.54±0.02	0.92±0.05	0.79±0.03	2.99±0.09	3.17±0.06
Al_2O_3	10.37±0.08	10.56±0.09	9.57±0.06	9.43±0.07	17.01±0.26	17.02±0.14	11.48±0.14	11.51±0.10
$SiO2$	69.53±0.21	70.40±0.26	78.33±0.12	78.68±0.27	59.18±0.30	58.93±0.24	52.65±0.17	53.28±0.21
P_2O_5	0.170±0.010	0.180±0.010	0.075±0.006	0.067±0.008	0.100±0.010	0.110±0.020	0.170±0.010	0.170±0.010
K_2O	2.29±0.06	2.28±0.02	2.47±0.05	2.50±0.03	0.98±0.03	1.08±0.02	2.09±0.04	2.07±0.02
CaO	1.63±0.05	1.73±0.03	0.81±0.04	0.74±0.03	0.17±0.04	0.17±0.01	11.47±0.10	11.48±0.06
TiO_2	0.75±0.02	0.75±0.01	0.84±0.03	0.83±0.01	1.56±0.04	1.61±0.04	0.64±0.02	0.63±0.01
MnO	0.077±0.002	0.079±0.002	0.070±0.002	0.069±0.002	0.051±0.002	0.052±0.001	0.089±0.003	0.085±0.002
Fe_2O_3	3.81±0.05	3.87±0.04	2.98±0.05	3.01±0.03	7.86±0.08	7.97±0.05	4.60±0.05	4.73±0.05
S	0.069±0.015	0.63±0.005	(0.04; 0.16)		0.040±0.010	0.040±0.005	0.040±0.010	0.040±0.005

Component	SDPS podzol sandy loam		SDO-2 marine sediment		SGH-1 carbonate background silt		SGHM-3 friable aluminosilicate deposit	
	Certified	XRF	Certified	XRF	Certified	XRF	Certified	XRF
Na_2O	0.51±0.03	0.53±0.01	4.03±0.04	4.03±0.03	0.53±0.02	0.56±0.02	0.61±0.04	0.65±0.04
MgO	0.13±0.05	0.11±0.01	4.67±0.08	4.34±0.09	6.06 ± 0.11	6.01±0.16	11.70±0.14	11.80±0.20
Al_2O_3	3.36±0.11	3.23±0.14	14.33±0.17	14.29±0.12	9.48 ± 0.14	9.32±0.07	5.03±0.10	5.06±0.04
$SiO2$	91.24±0.23	90.86±0.65	43.61±0.12	44.43±0.27	47.00 ± 0.20	46.68±0.29	25.07±0.29	25.24±0.21
P_2O_5	0.036±0.006	0.037±0.002	0.230±0.020	0.290±0.010	0.13 ±0.01	0.150±0.005	1.820±0.050	1.830±0.020
K_2O	1.23±0.03	1.31±0.02	1.36±0.02	1.37±0.03	2.26 ± 0.07	2.26±0.02	1.13±0.04	1.11±0.02
CaO	0.27±0.03	0.25±0.02	7.81±0.12	7.93±0.08	7.76 ± 0.10	7.77±0.08	17.76±0.22	17.99±0.31
TiO_2	0.29±0.01	0.25±0.01	2.32±0.06	2.30±0.03	0.50 ± 0.02	0.55±0.01	0.27±0.01	0.27±0.01
MnO	0.011±0.001	0.011±0.001	0.270±0.010	0.270±0.005	0.30 ± 0.01	0.300±0.005	0.500±0.030	0.500±0.010
Fe_2O_3	0.99±0.05	0.95±0.04	11.91±0.09	11.99±0.20	5.92 ± 0.04	5.94±0.04	10.59±0.20	10.92±0.20
S			(0.04; 0.16)		(0.037)		0.050±0.010	0.050±0.005

Component	SGH-3 terrigeneous background silt		SGH-5 anomalous silt		SGHM-1 friable carbonate-silicate deposit		SChT typical black earth soil	
	Certified	XRF	Certified	XRF	Certified	XRF	Certified	XRF
Na_2O	1.61±0.05	1.48±0.03	2.33±0.06	2.36±0.02	0.87±0.05	0.91±0.02	0.81±0.02	0.76±0.02
MgO	1.60±0.05	1.59±0.03	2.54±0.06	2.34±0.05	5.82±0.10	5.70±0.10	0.95±0.03	1.01±0.03
Al_2O_3	16.46±0.19	15.73±0.20	14.40±0.11	13.69±0.12	11.60±0.13	10.77±0.10	9.81±0.14	10.07±0.07
$SiO2$	60.54±0.20	59.42±0.22	60.85±0.14	60.95±0.23	45.59±0.29	46.90±0.60	71.49±0.27	72.35±0.25
P_2O_5	0.190±0.010	0.180±0.010	0.180±0.10	0.110±0.010	0.150±0.010	0.130±0.005	0.180±0.020	0.180±0.010
K_2O	2.43±0.08	2.45±0.03	3.56±0.09	3.39±0.03	2.96±0.07	2.90±0.03	2.42±0.04	2.49±0.02
CaO	0.41±0.03	0.48±0.03	2.95±0.05	2.87±0.04	7.05±0.20	6.81±0.07	1.60±0.05	1.79±0.20
TiO_2	0.98±0.03	0.96±0.01	0.62±0.01	0.58±0.01	0.63±0.04	0.68±0.01	0.74±0.03	0.73±0.01
MnO	0.130±0.010	0.140±0.005	0.087±0.003	0.086±0.002	0.073±0.004	0.082±0.002	0.079±0.002	0.081±0.002
Fe_2O_3	8.76±0.08	8.77±0.06	5.45±0.10	5.28±0.04	4.62±0.06	4.75±0.05	3.48±0.06	3.45±0.04
S	(0.027)		(0.10)		0.050±0.010	0.050±0.005	0.050±0.010	0.068±0.005

Component	GSS-4, limy-red soil		GSS-2, chestnut soil		GSS-5, yellow-red soil		GSD-10, stream sediment	
	Certified	XRF	Certified	XRF	Certified	XRF	Certified	XRF
Na_2O	0.11±0.01	0.12	1.62±0.02	1.65	0.122±0.009	<0.1	(0.04)	<0.1
MgO	0.49±0.02	0.60	1.04±0.02	1.05	0.61±0.02	0.67	0.12±0.02	0.20
Al_2O_3	23.45±0.11	25.92	10.31±0.05	10.17	21.68±0.09	24.46	2.84±0.04	3.35
$SiO2$	50.95±0.08	57.31	73.35±0.11	74.56	52.57±0.25	50.85	88.89±0.12	89.42
P_2O_5	0.159±0.003	0.218	0.102±0.002	0.085	0.089±0.004	0.102	0.062±0.002	0.054
K_2O	1.03±0.03	1.17	2.54±0.02	2.39	1.50±0.02	1.57	0.125±0.007	0.27
CaO	0.26±0.02	0.36	2.36±0.02	2.37	(0.095)	0.01	0.70±0.02	0.71
TiO_2	1.801±0.027	1.969	0.452±0.005	0.415	1.049±0.015	1.121	0.212±0.005	0.18
MnO	0.183±0.004	0.202	0.066±0.001	0.061	0.176±0.004	0.175	0.130±0.002	0.12
Fe_2O_3	10.30±0.05	11.16	3.52±0.03	3.33	12.62±0.08	13.34	3.86±0.04	3.38
S	0.018±0.003	<0.02	0.021±0.003	<0.02	0.041±0.004	0.038	0.009±0.00	<0.02

Table 1. Comparison of XRF results and certified and recommend values for Russia national certified and Chinese reference standards (%), respectively

3. A case study of the XRF determination of Na, Mg, Al, Si, P, S, Cl, K, Ca, Mn, Fe, Ni, Cu, Zn, Rb, Sr and Zr in dry powder of fish muscle tissue

The study of fundamental relations between natural constituents of aquatic ecosystems, as well as multi-purpose ecological investigations, focused on assessment of environment state and its protection from man-made impact, necessitates development of targeted analytical methods. Notice that when investigating aquatic ecosystems in the context of indicative ecology attention was so given to Hg, and to a lesser extent to Cd and Pb behavior (Nemova, 2005). The list of other elements to be examined was very constrained (Moiseenko, 2009) because multielement instrumental techniques became available only in the late the 20th century (Vetrov & Kuznetsova, 1997). At the moment diverse instrumental techniques, e.g. AAA, XRF, AES, NAA, AES ICP and MS ICP, are widely applied in the investigations of living matter of aquatic ecosystems (Moiseenko, 2009; Tolgyessy & Klehr, 1987), each method having specific limitations in terms of detection limit, selectivity and expressness (Vetrov & Kuznetsova, 1997; Kuznetsova et al., 2002).

The living matter of aquatic ecosystems is the medium uncommon for XRF, because of specific bioorganic composition; high and low contents of water and total mineral components, correspondingly, and in addition, lack of proper multicomponent certified standard materials (CRM) of both of national and international production. Utilization of artificial mixtures for calibrating and evaluating the accuracy by the «introduced-defined» method is restrained by the lack of inert material of required purity, as well uncertainty of modeling mixtures compositions. This part of paper reports information on nondestructive XRF procedure to determine the contents of elements Na, Mg, Al, Si, P, S, Cl, K, Ca, Mn, Fe, Ni, Cu, Zn, Rb, Sr and Zr in dry powder of fish muscle tissue.

3.1 Samples and their preparation

A series of 60 emitters were produced from certified reference material of composition of Baikal perch muscle tissues BOk-2 (CRM No.9055-2008 (BOk-2)) (CATALOGUE, 2009) to be employed in the experiment. Three emitters were made of the materials of twenty sealed and labeled polyethylene jars. Besides, our objective was muscle tissue powders of omul, golyan and river perch (group I), the same as perch and plotva (group II). Fishes of group I were collected in the Chivyrkuy Bay of Baikal Lake, while those of group II were caught in various parts of the Baikal and Bratsk man-made water reservoir. The sampling sites differed in the rate of technogenic pollution.

The muscle tissues of group II fishes were lyophilized to a steady weight using a Labconco lyophilizer (method 1). The muscle tissue of omul, river perch and golyan (fishes of group I) were slowly dried to a steady weight on the water bath at $T=60\pm3°C$ (method 2). The emitters weighing 4 g were pressed under 4 tons pressure in the mold heated to $T=38\pm0.1°C$ (Gunicheva et al., 2005). The temperature was regulated by TRM-101 thermostat manufactured by TERMIK Co. in Moscow. The emitting layer for the analytical lines of elements Na, Mg, Al, Si, P, S, Cl, K, Ca, Mn, Fe, Ni, Cu and Zn is thick, whereas for the lines of elements Rb, Sr, Zr and Rh it is intermediate. The uncertainty of the thickness of emitting layer is assigned to weighing mass 4 g on weights BP 61S Sartorius, Max 61, d=0.1 mg.

3.2 XRF equipment

The intensities of analytical lines and background were measured in vacuum by the X-ray spectrometer with wave dispersion S4 Pioneer (Bruker Firm, Germany). Temperature in a vacuum cell is equal to 38° C. The conditions for excitation and registration of x-ray fluorescence and background are listed in Table 2.

X-ray tube with Rh–target, Be–window, 0.0075 cm thickness, and incidence angle of the primary and exit one of the secondary radiations are equal to 63 ° and 45 °, accordingly.									
Ana-lyte	2θ, °		Crystal	Detec-tor	Time, s		Voltage, kV	Current, mA	Collimator, °
	$K\alpha_1$-	Back-ground							
Na	24.90	23.87 / 25.89	OVO-55	PC	100	50 / 50	30	40	0.46
Mg	20.58	21.96	OVO-55	PC	30	30	30	40	0.46
Al	144.61	145.82	PET	PC	60	60	30	40	0.23
Si	108.99	109.78	PET	PC	30	30	30	40	0.23
P	89.43	91.40	PET	PC	10	10	30	40	0.46
S	75.75	74.77	PET	PC	10	10	30	40	0.46
Cl	65.41	66.87	PET	PC	10	10	30	40	0.46
K	136.67	139.54	LiF(200)	PC	10	10	30	40	0.46
Ca	113.11	115.17	LiF(200)	PC	10	10	30	40	0.46
Mn	62.97	62.26 / 63.72	LiF(200)	PC	30	15 / 15	50	40	0.23
Fe	57.52	58.28	LiF(200)	PC	30	30	50	40	0.23
Ni	48.66	48.17 / 49.08	LiF(200)	CC	30	15 / 15	50	40	0.23
Cu	45.04	44.40 / 45.71	LiF(200)	CC	30	15 / 15	50	40	0.23
Zn	41.75	40.98 / 42.53	LiF(200)	CC	30	15 / 15	50	40	0.23
Rb	26.61	26.08 / 27.23	LiF(200)	CC	30	15 / 15	50	40	0.23
Sr	25.14	24.61 / 25.62	LiF(200)	CC	30	15 / 15	50	40	0.23
Zr	22.51	21.74 / 23.21	LiF(200)	CC	30	15 / 15	50	40	0.23
Rh	18.47	-	LiF(200)	CC	10	10	50	40	0.23

Table 2. Conditions of excitation and registration of XRF by S4 Pioneer spectrometer.

The emitter was being measured for no more than 19 minutes. After measuring 9 emitters (in ~6 hours) "the reference emitter" made from certified reference material of Tr-1 (see

Table 3) was measured in order to control the equipment drift. It permitted to use both absolute and relative intensities. For estimating the long-time stability of emitter from dry powder of fish muscle tissue all measurements were carried out during a year. The measurements for every emitter were executed $6 < n < 12$ times, in total 480.

Sample No	Reference Material	Producers
1	Tea (GSV-4)	Institute of Geophysics. & Geochem. Exploration (IGGE), Heibei, China
2	Leaf of birch (Lb-1)	Siberian Branch of Russian Academy of Sciences, Institute of Geochemistry, Irkutsk, Russia.
3	Mixture of meadow herbs (Tr-1)	Siberian Branch of Russian Academy of Sciences, Institute of Geochemistry, Irkutsk, Russia.
4	Canadian pondweed (Ek-1)	Siberian Branch of Russian Academy of Sciences, Institute of Geochemistry, Irkutsk, Russia.
5	Baikal perch tissue (BOk-2)	Siberian Branch of Russian Academy of Sciences, Institute of Geochemistry, Irkutsk, Russia.
6	Potatoes tuber (SBMK-02)	Central Institute of Agrochemical Service of Agriculture and Sverdlovsk Branch VNIIM
7	Wheat grain (SBMP-02)	Central Institute of Agrochemical Service of Agriculture and Sverdlovsk Branch VNIIM
8	Cereal herb mix (SBMT-02)	Central Institute of Agrochemical Service of Agriculture and Sverdlovsk Branch VNIIM
9	Milk Powder IAEA – 153	Report: IAEA/AL/010 Australia
10	Milk Powder IAEA A11	Report: IAEA/AL/010 Australia
11	Microcrystal cellulose	Sigmacell Cellulose, Type 50, S5504-1KG
12	Aminoethanole acid	analyzed by ICP-OES using certified technique
13	L-asparagine	analyzed by ICP-OES using certified technique
14-16	Milk-based infant formulas	International Nutrition Co, Denmark

Table 3. Specifications of certified reference materials and samples

3.3 Content computation

The key points of procedure: generation of calibrating samples set, taking into account their physical and chemical properties; optimization of approximation capacity of calibration functions due to a proper selection of regression equation and regression approach to determine the parameters of calibration function. The certified reference materials and samples are tabulated in Table 3. Characteristics of calibration collection are given in Table 4. The influence of inadequacy of composition of organic matrix of the certified reference materials set and bioorganic matrix of fish muscle tissue on the XRF results of fish tissues was not studied yet. Only the principal XRF fitness was estimated. Admitting a rough similarity of bioorganic compositions of dry residue of cow milk and fish muscle tissue, it is believed that the systematic error due to this effect will not be over 8 %

(Gunicheva, 2010). The calibration functions have been selected out of the calibrations implemented by the software of X-ray spectrometer S4 («SPECTRAPLUS», 2002). The approach of alpha coefficients was applied for correcting the matrix effects when measuring elements Na, Mg, Al, Si, P, S, Cl, K and Ca:

$$C_i = m_iI_i * (1+\sum \alpha_{ij}I_j), \tag{1}$$

where: (a) C_i is the concentration of analyte i; (b) I_i is intensity of its analytical line corrected for the background; (c) m_i is slope of calibration plot; (d) I_j is intensity of matrix element analytical lines corrected for the background; (e) α_{ij} is the value of the corresponding alpha coefficient calculated by the linear regression equation. The contents of Mn, Fe, Ni, Cu, Zn, Rb, Sr and Zr were analyzed by the background standard method. A characteristic line of the x-ray tube anode (see Table 2), incoherently scattered from sample, was the standard. The concentration of analyte i was calculated by the equation:

$$C_i = m_i* (I_i + K_i)/I_j, \tag{2}$$

where: (a) I_j is pure intensity of comparison line; (b) K_i is expression of intensity correction, (c) m_i is slope of calibration plot. Intensity I_j is proportional to $1/(1+M)$, where M is the coefficient for correcting matrix effects. The parameters of calibration functions (1-2) were optimized by regression approach («SPECTRAPLUS», 2002).

Analyte	Interval, ng/g	RM quantity
Na	40 - 6900	9
Mg	40- 4400	10
Al	20 - 3000	8
Si	2- 5450	9
P	80 - 3600	11
S	1000 - 3600	11
Cl	200 - 8400	8
K	20 - 23900	11
Ca	10 - 16200	11
Mn	0.2 - 1240	10
Fe	2.5 - 990	8
Ni	0.7 – 5.8	8
Cu	0.4 – 17.3	10
Zn	2 - 94	9
Rb	3.5 - 74	10
Sr	2.2 - 72	10
Zr	0.2 – 5.5	8

Table 4. Characteristics of calibration collection

3.4 The temporal trends of X-ray fluorescence intensities

The regression equations of temporal trends for x-ray fluorescence intensities, both absolute and relative ones, for the elements to be determined are provided in Table 5.

Parameters of equations are presented as the range of their magnitudes, obtained for the entire series of emitters. Both absolute and relative intensities of all elements are expressed as $R^2_{exp} < r_{xy}$. It is proposed to accept: (a) absence of paired correlation and (b) contribution of temporal change is small in comparison with the discrepancy in values of a_0 and a_1 for the emitters pressed from various polyethylene jar materials. This fact is no surprise, since material of CRM BOk-2, as any biological medium, is *a priori* natural non-equilibrium system (Vernadsky 1978; Marchenko, 2003). The data in Table 5 disclose stress of processes proceeding within substance when converted in powder state, x-ray irradiation and effects of increased temperature and vacuum, set by instrument parameters of S4 Pioneer («SPECTRAPLUS», 2002). The influences of the specified processes on the accuracy of XRF data on fish tissue are still to be properly considered. The further study would require a thorough planning and implementing with fish tissue material of a set mass and fresh sample preparation.

	Absolute intensities	R^2_{exp}	Relative intensities	R^2_{exp}
Na	y=-0.001x+(28÷45)	0.3÷0.5	y=-(0.001÷0.002)x+(25÷63)	0.0÷0.2
Mg	y=-(0.001÷0.003)x+(43÷82)	0.3÷0.6	y=-(0.001÷0.002)x+(40÷72)	0.1÷0.6
Al	y=-(0.000÷0.001)x+(12÷45)	0.3÷0.5	y=-(0.000÷0.001)x+(14÷36)	0.0÷0.1
Si	y=0.001x+(28÷45)	0.3÷0.5	y=-(0.000÷0.001)x+(12÷46)	0.0÷0.2
P	y=-(0.01÷0.03)x+(463÷980)	0.2÷0.5	y=-(0.000÷0.001)x+(15÷40)	0.0÷0.1
S	y=-(0.03÷0.07)x+(1087 ÷2155)	0.2÷0.6	y=-(0.000÷0.001)x+(14÷33)	0.1÷0.3
Cl	y=-(0.004÷0.010)x+(183÷348)	0.1÷0.5	y=-(0.000÷0.001)x+(17÷28)	0.1÷0.3
K	y=-(0.05÷0.11)x+(0.11÷0.43)	0.1÷0.4	y=-(0.000÷0.001)x+(18÷32)	0.1÷0.2
Ca	y=-(0.004÷0.007)x+(169÷288)	0.3÷0.5	y=-(0.00÷0.01)x+(13÷25)	0.1
Mn	y=-(3E-06÷7E-05)x+(1.1÷1.9)	0.2÷0.5	y=(0.6÷0.7)x+(0.24÷0.38)	0.5÷0.6
Fe	y=-0.001x+(47÷551)	0.5	y=-0.001x+(21÷58)	0.5
Ni	y=-9E-05x+(3.6÷4.8)	0.2÷0.3	y=-(0.000÷0.001)x+(16÷37)	0.0÷0.1
Cu	y=-0.001x+(30÷53)	0.3÷0.2	y=1.0x+(0.01÷0.03)	0.4
Zr	y=-1E-05x+(0.1÷2.8)	0.0÷0.3	y=-(0.000÷0.001)x+(15÷40)	0.2÷0.5
Sr	y=-1E-05x+(2÷10)	0.02	y=-(0.000÷0.001)x+(14÷24)	0.2÷0.5
Rh	y=-(0.000÷0.001)x+(2.6÷2.8)	0.0÷0.4	y=-(0.000÷0.001)x+(5÷9)	0.3÷0.5

Table 5. Regression equations of temporal trends; * r_{xy} = 0,537 for $p < 0.01$

3.5 Selection of intensities for concentration computation

The series of absolute and relative intensities for the emitters were compared by two-factor analysis of variance with different dispersions. The results are given in Table 6.

Analyte	Trend	Average	Dispersion	n	t_{exp}	Resume
Na	abs.	0.4413	0.0325	20	1.75	By absolute intensities
	rel.	0.3229	0.0577	20		
Mg	abs.	0.4905	0.0171	20	2.08	By absolute intensities
	rel.	0.3760	0.0424	20		
Al	abs.	0.2177	0.022	20	1.70	By absolute intensities
	rel.	0.1400	0.0187	20		
Si	abs.	0.3236	0.0212	20	1.03	Both schemes are comparable
	rel.	0.2679	0.0354	20		
P	abs.	0.3874	0.0180	19	1.45	Both schemes are comparable
	rel.	0.3064	0.0423	20		
S	abs.	0.4072	0.0177	19	1.56	Both schemes are comparable
	rel.	0.3214	0.0404	20		
Cl	abs.	0.3003	0.0161	20	2.13	By absolute intensities
	rel.	0.1941	0.0328	20		
K	abs.	0.2822	0.0104	19	2.19	By absolute intensities
	rel.	0.1871	0.0259	20		
Ca	abs.	0.2270	0.0159	20	0.74	Both schemes are comparable
	rel.	0.1921	0.0267	20		
Mn	abs.	0.2776	0.0622	20	-0.18	Both schemes are comparable
	rel.	0.2922	0.0701	20		
Fe	abs.	0.5318	0.0427	20	-0.24	Both schemes are comparable
	rel.	0.5456	0.0282	20		
Ni	abs.	0.2250	0.0152	20	1.45	Both schemes are comparable
	rel.	0.1635	0.0207	20		
Cu	abs.	0.3976	0.0726	20	0.34	Both schemes are comparable
	rel.	0.3574	0.0849	20		
Zn	abs.	0.2205	0.0198	20	-0.36	Both schemes are comparable
	rel.	0.2400	0.0452	20		
Rb	abs.	0.0215	0.0014	18	-3.21	By relative intensities
	rel.	0.1001	0.0104	20		
Zr	abs.	0.1272	0.0368	20	-1.73	By relative intensities
	rel.	0.2324	0.0374	20		
Sr	abs.	0.0653	0.0058	20	-1.82	By relative intensities
	rel.	0.1534	0.0407	20		
Rh	abs.	0.1577	0.06786	20	0.03	Both schemes are comparable
	rel.	0.1554	0.0377	20		
t critical one-way $p=0.01$				1.69		abs. –absolute intensities
t critical two-way $p=0.01$				2.03		rel. –relative intensities

Table 6. Comparison of absolute and relative intensities.

The empirical values of t - Student coefficients t_{exp} are more tabular t_{tab} for elements Na, Mg, Al, Cl, K (set 1) and less for Rb, Zr and Sr (set 2). Therefore to compute concentration for elements

of set 1 the absolute intensities were used and for set 2 elements these were relative ones. For
elements Si, P, S, Ca, Mn, Fe, Cu и Zn selection of the intensities for concentration computation
is non-critical and XRF analysis results are comparable when using both intensity sires.

3.6 The metrological characteristics of the XRF procedure

Constituents of random error were assessed by three-factor analysis of variance. The values
characterizing convergence error of intensity measurement $S_{r.c}$, the error of emitter
preparation and its setting up in the holder S_{rp} and total intralaboratory XRF accuracy error
$S_{r,tot}$ are summarized in Table 7. It also provides the ranges $N_{min} - N_{max}$ and the maximum
values of count statistics error $1/\sqrt{N}$ (N denotes the number of counts) to ease
understanding. N_{max}/N_{min} ratios are changed from 2.5 for Zr to ~1.1 for elements Mg, Al, Si,
P, K, Cu, Zn and Rb. In column of $S_{r,tot}$ values the brackets enclose the estimations
computed for relative intensities. It is evident that for only Ca and S statistics errors are the
dominant contributions into evaluation of total intralaboratory XRF random error. For the
rest of analytes the effects are not so simple. For elements Si, Cl, Mn, Ni, Cu, Zn, Rb, Sr and
Zr values of $S_{r,tot}$ are mainly caused by error of intensity measurement. For analytes Na,
Mg, Al. P, K, Fe and Rh the errors of emitter preparation and its setting up in the holder are
maximal signified, i.e. behavior of each analyte is unique and requires careful consideration.

Ana-lyte	N_{min}-N_{max}, counts	$1/\sqrt{N}$	$S_{r.c}$	S_{rp}	$S_{r,tot}$	$t_{2,1\ exp}$	$t_{3,2\ exp}$	C_{min}, ppm
Na	83.6÷184.8	0.11	2.26	10.14	9.23 (13.61)	47.90	2.00	*
Mg	0.76÷0.82	1.15	0.45	1.36	1.19 (1.21)	19.25	2.32	*
Al	0.854÷0.885	1.08	0.34	0.81	0.74 (0.93)	12.32	1.99	*
Si	0.233÷0.242	2.07	8.92	n.s.	8.45 (7.60)	1.17	0.08	*
P	1.482÷1.567	0.82	0.31	0.93	0.81 (0.97)	19.19	1.45	*
S	0.13÷0.15	2.77	0.48	1.97	1.67 (2.36)	34.03	2.77	*
Cl	0.0291÷0.0448	5.86	12.60	n.s..	10.34 (9.04)	0.50	1.22	3.2
K	1.119÷1.142	0.95	1.06	2.24	2.08 (2.77)	10.00	2.37	5.0
Ca	0.0134÷0.0216	8.64	4.61	9.06	8.55 (10.47)	8.71	1.64	4.3
Mn	0.0315÷0.0448	5.63	12.60	n.s.	10.34 (8.42)	0.50	1.22	0.9 (2.8)
Fe	0.922÷1.170	1.04	4.50	4.81	5.71 (6.13)	3.29	0.94	1.3 (2.6)
Ni	0.173÷0.218	2.40	7.67	n.s.	7.01 (6.98)	0.98	0.05	0.4 (0.8)
Cu	1.144÷1.257	0.94	2.03	n.s.	1.67 (1.76)	0.53	1.92	1.4 (1.4)
Zn	2.566÷2.775	0.62	1.32	n.s.	1.16 (1.20)	0.8	1.25	0.4 (1.1)
Rb	4.79÷5.11	0.46	1.24	n.s.	1.15 (1.32)	1.06	2.50	0.5 (1.6)
Sr	0.435÷0.590	1.52	8.71	n.s.	7.15 (7.25)	0.51	1.65	0.3 (1.8)
Zr	0.095÷0.238	3.24	21.74	n.s.	20.79 (19.93)	1.22	0.64	0.3
Rh	1.667÷1.988	0.77	1.97	2.68	2.85 (3.12)	4.72	1.09	
$t_{2,1}(0,01,40,60)_{tab.}=3.99$ $t_{3,2}(0,01,19,40)_{tab.}=2.03$	n.s.- insignificant; * stands for the elements with limits of contents being essentially beyond 10σ.							

Table 7. Estimates, % of random error components

Column of C_{min} presents detection limits, calculated with 3σ-criterion using the results of 20 measurements of emitters produced from powder of L- asparagine, aminoethanole acid, cellulose and reference samples OM-1 and OK-1 (muscle tissues of the Baikal omul and perch). They vary from $(3\div9) \times 10^{-6}$ for elements of Mn, Ni, Zr, Rb, Sr and Zn to $(2\div5) \times 10^{-5}$ for Cl, K, Ca, Fe and Cu, which is to say that XRF data are acceptable to investigate the living matter of aquatic ecosystems.

The detection limits were not given for elements of Na, Mg, Al, Si, P and S, because their contents in fish tissues exceed substantially the limits of quantitative determination (10σ). The brackets enclose detection limits from reference (Gunicheva et al., 2005). The larger magnitudes for the latter are due to the difference in bioorganic and organic compositions of dry powder of fish muscle tissues and plant materials.

3.7 The accuracy of XRF data

The classical assessment of accuracy of XRF data on the fish tissues would be impossible because of unavailability of proper certified reference materials and reference samples (ISO, 1994). Therefore, the XRF results for muscule tissues of various fishes have been compared with similar literature data. These data (Vetrov & Kuznetsova, 1997; Leonova, 2004; Moore & Ramamurti, 1987; Grosheva et al., 2000) are given for the fishes, collected in the southern and middle Baikal, Selenga River estuary, Angara River in the environments of Bratsk and Ust'- Ilimsk man-made water reservoirs (i.e. for the sites of CRM BOk-2 sampling), and were acquired by the authors through span 1987-2005 by different instrumental techniques.

Table 8 presents their types and sample preparation described in the references. It indicates that digestion is prevailing in sample preparation, when investigating the living matter of aquatic ecosystems.

Reference	Type of benefic organism	Sampling site	Instrumental technique	Sample preparation
Leonova, 2004	Golyan, perch, plotva, omul	Chivyrkuy Bay of Baikal Lake	Atomic emission spectrometry (AES)	Digestion
Leonova & Bychinskiy, 1998	Perch	Bratsk man-made water reservoir, Lake Baikal	AES with evaporating sample powders in canal of arc graphite electrode	Digestion
Grosheva et al., 2000	Perch, plotva	Lake Baikal	Substoichiometric isotope dilution	Lyophlization
Moore & Ramamurti., 1987	Perch	The upper stream of river Ob' in site of town Barnaul water intake	Atomic absorption spectrometry (AAS)	Solubilizing
Vetrov & Kuznetsova, 1997	Omul, plotva, perch	Lake Baikal	AES	Digestion

Table 8. Details of instrumental techniques and sample preparation used in references

Comparison is presented in Tables 9-10. At Table 9 the column of analytes shows the elements certified for reference material BOk-2 by bold print; italics type designate for recommended

Application of Nondestructive X-Ray Fluorescence Method (XRF) in Soils, Friable and Marine Sediments and Ecological Materials

143

Analyte	CRM BOk-2	Bratsk man-made water reservoir		Southern and Middle parts of the Baikal					
		XRF*	Leonova &, Bychinskiy, 1998; Moore & Ramamurti, 1987	XRF*	XRF$^+$	Grosheva et al., 2000	Vetrov & Kuznetsova, 1997	Leonova, 2004; Ciesielski et al, 2000	Ciesielski et al., 2005
Al	28±18	10.3±2.6		8.6±1.6	31±3		17±8	3.8±0.9	
Fe	53±11	25.9±7.2	10.0	10.97±3.83	33.4±2.4	61.87±7.17	41±21	57.2±3.9	672
K	15900±700	14002±589		15476±307	15653±88				
Mn	1.66±0.24	0.60±0.35	10.0	1.11±0.35	1.09±0.18	2.24±0.22	0.64±0.06	3.2±0.6	
Ni	0.42±0.27	1.09±0.05	0.2	1.07±0.06	0.54±0.16		0.26±0.13	2.24±0.18	
Rb	21.9±4.3	19.3±2.4		33.9±1.2	21.9±0.4				
S	11000 ± 2000	9304 ±259		8723 ± 226	10412±54				2.0
Sr	2.8 ± 0.3	3.5 ± 0.8		2.4 ± 0.3	2.7 ± 0..3			194±31.1	
Zn	23.0±1.2	32.4±2.6	18.2÷37.2	27.3±1.4	27.5±4.7	9.29 ± 1.25	24±6	113.4±17.1	80
Ca	1720±250	1125±174		859±114	1452±34			4911±119	
Mg	1040±110	1066±66		1325±54	1261±29			1484±103	
Na	2770±90	4836±648		3438±258	2778±42			5444±495	
P	9500±500	7946±293		9329±211	9289±63			9977±12.37	
Cl	2800 ± 200	2499±404		2074 ± 157	2494± 24				
Si		35.9±6.7		33.7 ±15.3	106±8			6.3±0.1	
Br	49 ± 5	23.3±5.7		99±6	55±0.2				

Table 9. Contents of analytes in perch muscle tissues (dry weight, ng/g

Ana-lyte	Golyan		Omul		Perch		Plotva	
	XRF*	Leonova, 2004	XRF*	Leonova, 2004	XRF*	Leonova, 2004	XRF+	Leonova, 2004
Na	6184±182		4840±118	3166±171	4229±163	2600±204	3000±500	5750±531
Mg	820±21	32,70	1108±23	2700±307	1454±50	1933±1160	1250±80	2750
Al	43±3	2,43	41±3	105±51	26±4	0,9±0,3	11±4	46
Si	145±4		159±4	968±354	82±10	7,5±1,8	33±9	550±437
P	11328±317		8732±168	6333±341	8618±278	3500±341	8710±390	2500±625
S	6762±110		6182±73		9176±270		8310±340	
Cl	3032±94		3492±125		2289±96		1660±370	
K	8730±150		14052±210		13620±336		14800±400	
Ca	13100±210		740±10	6333±341	1268±30	1000±546	830±80	2500
Mn	3,65±0,67	0,22	1,34±0,42	3,3±1,8	0,66±0,22	2,4±1,0	1,03±0,08	
Fe	68±1	3,14	35±8	146±20,5	62±1	55,0±13,6	26±6	34,5
Ni	0,80±0,02		1,05±0,07	0,39±0,08	0,92±0,04	0,3±0,1	1,08±0,11	1,7
Zn	22,30±0,71	0,72	24,4±0,8	12,6±0,7	13,16±0,05	27,3±9,5	39,7±6,6	47,5
Rb	1,6±0,1		26,1±1,4		2,4±0,2		9,1±1,1	
Sr	11,7±0,1		3,2±0,2	52±2,7	0,5±0,1		3,1±1,1	
Br	11,4±0,2		55,1±1,5		8,9±0,2		29,3±12,1	

Table 10. Contents of analytes in muscle tissues of different fish kinds, (dry weight, ng/g) The sampling site is the Chivyrkuy Bay of Baikal Lake

value; * denotes for the median of sampling and empty cell shows data absence. The data of the XRF columns have been gained by statistic treatment of no less than ten samples (free emitters for each). It is safe to say that the XRF information bulk for fish tissues is more excessive in comparison with the referenced information. It should be noted that information on content of Br, Cl, S, K, Cs, Sr and Sc in fishes of Baikal region is entirely absent in the references. Contents for the other elements belong in concentration intervals given by the other authors.

Microelement contents in perch tissues from Novosibirsk man-made reservoir (river Ob') (Leonova, 2004) are also comparable with the data for BOk-2. Notice that among the references, the only the data of (Leonova, 2004) is close to XRF results being conformable. Considering features of used analytical techniques some discrepancy being visible to the human eye is existent. Nevertheless the observed ranges of element contents are beyond methodical errors. They are indicative of fish tissue composition dependence on situation of aquatic ecosystems. We emphasize that irrespective of the difference in analytical techniques, the levels of concentrations and a series of decreasing element contents P> Na> Ca > Mg > Zn > Mn are fairly similar in all data.

As to Table 10 for all fish kinds in general consistency of data is enough apparent regardless of the fact that instrumental techniques and sample preparations are not identical.

Table 11 gives some relationships between the XRF concentrations of some elements for muscle tissues of plotva and perch fish, collected in the southern and middle Baikal and Bratsk man-made water reservoir, as well as omul. Statistically significant interrelationships are given by bold print. These correlations are conformable to those, represented in the liver of Baikal seal (Ciesielski et al., 2006). They reflect not only techniques used when preparing samples and conditions of ecosystem in the sampling sites, but also fish trivial heredity. They can give usable information to interpret environment impact and element interactions with numerous factors, both biotic and abiotic. These correlations also demonstrate that the XRF data may be utilized as the efficient and sensitive indicator of changing element constituents of geochemical background.

Analyte pair	Bratsk man-made water reservoir		Various parts of the Baikal		the Chivyrkuy Bay
	perch	plotva	perch	plotva	omul
Mn-Br	0.05	-0.28	-0.46	**-0.91**	-0.13
Fe-Br	-0.43	-0.54	**-0.80**	**-0.93**	0.13
Zn-Br	**-0.76**	**-0.65**	**-0.94**	**-0.99**	**-0.58**
Rb-Br	**0.97**	**0.99**	**0.92**	**1.00**	0.16
Sr-Br	-0.47	**0.92**	-0.38	**-0.62**	-0.15
Zn - Rb	**-0.62**	**-0.57**	**-0.65**	**-0.97**	0.47
Zn - Sr	**0.84**	-0.40	-0.09	0.36	-0.34
Rb - Sr	-0.31	**0.77**	0.41	-0.49	-0.24
Fe - Zn	0.12	**0.84**	**0.73**	**0.66**	0.39
Zn - Al	0.36	0.28	**0.91**	-0.31	-0.11
Mn - Fe	-0.32	0.25	0.09	0.26	-0.41
Fe - Ni	**-0.56**	-0.18	-0.09	0.46	0.20
Na - Cl	**0.97**	**0.98**	**0.95**	**0.62**	**0.95**
Na - K	**-0.91**	0.48	0.53	-0.06	**0.96**
Mg - Ca	0.35	**0.91**	**0.93**	**-0.57**	**0.90**
S - P	-0.06	**0.93**	**0.90**	0.22	-0.05
S - Cl	**0.60**	**0.74**	**0.82**	0.33	-0.19
Cl - P	**-0.67**	**0.86**	0.00	**0.89**	**0.99**
Ca - P	0.34	**0.77**	-0.49	**0.74**	**0.93**
Al - Ca	**-0.53**	0.06	0.04	**-0.61**	0.48
Al - Na	**-0.90**	-0.12	-0.10	**-0.82**	**0.53**
Al - Si	**0.77**	0.49	-0.17	**-0.54**	0.41
Al - Sr	-0.20	0.10	-0.17	-0.48	-0.27
Ca – Na	0.48	**0.87**	0.02	**0.89**	**0.81**
Ca – Si	**-0.58**	**0.56**	-0.09	**0.96**	**0.80**
Ca – Sr	**0.94**	**0.95**	-0.09	**0.90**	0.02
Fe - K	**0.70**	**-0.73**	0.18	-0.30	0.32
Fe - Mg	**-0.62**	0.05	-0.12	**-0.87**	0.24
Fe - P	0.08	**-0.51**	-0.28	**-0.73**	0.30

Table 11. Some element correlations

4. Conclusion

Environmental problems and the modelling cycle of major and minor elements in soil-plant systems under natural conditions and in response to man's activities require the extensive analyses. Most of the analytical problems that occur are simple to solve by XRF spectrometry. This technique provides accurate analyses of rocks and materials which may be homogenized by fusion with an appropriate flux. Such an approach as a rule is impossible for XRF analysis of materials which are abundantly supplied with organic constituents. Its biochemical complexity inhibits homogenization and, as a result, does not allow the extension of this procedure.

We have demonstrated that the use of nondestructive XRF method ideally suits the quantitative determination of Na, Mg, Al, Si, P, S, Cl, K, Ca, Mn, Fe, Ni, Cu, Zn, Rb, Sr and Zr contents in dry powders of muscle fish tissues. The regression equations of temporal trends for x-ray fluorescence intensities, both absolute and relative ones, for the elements were determined and shown that contribution of temporal change is small in comparison with the discrepancy in values of a_0 and a_1 for the emitters pressed from various polyethylene jar materials. This fact is not subitaneous because of material of CRM BOk-2, as any biological medium, is *a priori* natural non-equilibrium system. This imposes the strict initial conditions of the similarity: means of sampling and treatment.

Constituents of random error were assessed by three-factor analysis of variance. It is evident that behavior of each analyte is unique and requires careful consideration and tracing.

The quantitative analyses indicate that in various environmental situations the samples of all kinds of fish contain enumerated elements in different amounts and, therefore, they confirm validity of using fishes as indicating metal contamination. The metrological parameters of the technique allowed the sources of the errors to be identified, and the issues of further investigations to be projected.

The tendencies identified with the XRF results on the dry powders of fish muscle tissues do not contradict the features recognized in toxicology of aquatic ecosystems and environmental biogeochemistry. They disclose the potential of their utilization for multi-purpose interpretations in environmental monitoring of freshwater ecosystems.

5. References

Arnautov N. V. (1987). *Reference Samples of Natural Media Composition.*Procedure Recammendations. Novosibirsk. 99 p.

Bock R. (1972) *Digestion methods in analytical chemistry.* Verlag Chemie GmbH, Weinheim/Bergstr. 432p.

Bolt G. H. and Bruggenwert M. G. M. (1976). *Soil Chemistry. A. Basic Elements*, 281 p. Elsevier, Amsterdam

CATALOGUE of reference materials of composition of natural and technogenic media. Irkutsk, (2009). http: / www. igc.irk.ru

Ciesielski T., Pastukhov M. V., Fodor P., Bertenyi Z., Namiesrnik J., Szefer P. (2006). Relationship and bioaccumulation of chemical elements in the Baikal seal (*Phoca sibirica*) *Environmental Pollution*, V. 139, no 2, 372-384.

Greenland D. J. and Hayes M. H. B. (1978). (Eds). *The Chemistry of Soil Constituents*, p. 469.
    Wiley, New York.

Grosheva E.I., Voronskaay G.N., Pastukhov M.V. (2000), Trace element bioavailability in
    Lake Baikal. *Aquatic Ecosystem Health and Management*, 3, 229-234.

Gunicheva T.N., Pashkova G.V., Chuparina E.V. (2005), Results on hot pressing applicability
    for non-destructive XRF of plants. *Analytics and Control*, 9, 273-279.

Gunicheva T.N. (2010). Advisability of X-ray fluorescence analysis of dry residue of cow
    milk applied to monitor environment. *X-Ray Spectrometry*, 39, 22-27

INTERNATIONAL REFERENCE MATERIALS A Compilation of Currently Certified or
    Accepted Concentrations LOS ALAMOS NATIONAL LABORATORY *Operated by
    the University of California for the US Department of Energy Copyright © UC 2000 -
    Disclaimer* http://www.geostandards.lanl.gov/MaterialsByNumber/htm

ISO 5725 -1 — 1994. Accuracy (trueness and precision) of measurement methods and results.
    Part 1. General principles and definitions.

ISO 5725 -2 — 1994. Accuracy (trueness and precision) of measurement methods and results.
    Part 2. Basic method for the determination of repeatability and reproducibility of a
    standard measurement method.

ISO 5725 -3 — 1994. Accuracy (trueness and precision) of measurement methods and results.
    Part 3. Intermediate measures of the precision of a standard measurement method.

ISO 5725 -4 — 1994. Accuracy (trueness and precision) of measurement methods and results.
    Part 4. Basic methods for the determination of the trueness of a standard
    measurement method.

ISO 5725 - 5 — 1998. Accuracy (trueness and precision) of measurement methods and
    results. Part 5. Alternative methods for the determination of the precision of a
    standard measurement method.

Kabata-Pendias A. and Pendias H. (1986). *Trace Elements in Soils and Plants*, CRC Press, Boca
    Raton, FL. 439 p.

Kuznetsova A.I., Zarubina O.V. and Leonova G.A. (2002), Comparison of Zn, Cu, Pb, Ni, Cr,
    Sn, Mo concentrations in tissues of fish (roach and perch) from Lake Baikal and
    Bratsk Reservoir,. *Environmental Geochemistry and Health*, 24, 205-212.

Lindsay W. L. (1979). *Chemical Equilibria in Soils*, p. 449. Wiley-Interscience, New York.

Ostroumov G. V (1979). The Metrological Basics of Exploration of Rock, Ore and Mineral
    Chemical Composition. The Bowels of the Earth. Moscow (Ed.). 400 p.

Leonova G.A., Bychinskiy V.A. (1998). Hydrobionts of the Bratsk water reservoir as the sites
    of heavy metal monitoring. *Water resources, 25, 603.*

Leonova G.A. (2004). Biogeochemical indication of natural and technogenic concentrations
    of chemical elements in components of aqua systems, exemplified by Siberian
    water reservoirs. *Electronic Journal «Explored in Russia»*, 197, 2196
    http://zhurnal.ape.replarn.ru/articles/2004/197.pdf.

Marchenko E.D. (2003). *Memory of globe experience.* St. Petersburg, Author Center
    "RADATS", 376 p

Moiseenko T.I. (2009). *Aqua toxicology: Theoretical and applied aspects.* Moscow, Nauka, 400p.
    ISBN 978-5-02-036166-9

Moore D.S., Ramamurti S. (1987). *Heavy metals in natural waters. Control and assessment of
    effect.* Moscow, Mir, 285 p.

Nemova N.N. (2005). *Biochemical effects of mercury accumulation in fish.* Moscow, Nauka. 200 p

Rudneva N.A. (2001), *Heavy metals and microelements in hydrobionts of the Baikal region*. Inst. Exper.Biology, SB RAS, Ulan-Ude. Publ. H. B.S.C. SB RAS, 134 p.

Schnitzer M. and Khan S. U. (1978). *Soil Organic Matter*, 319 p .. Elsevier, Amsterdam.

«SPECTRAPLUS» for users of spectrometer S4 EXPLORER. Karlsruhe: Bruker AXS Center. 2002.

Tolgyessy J. and Klehr E.H. (1987). *Nuclear methods of chemical analysis of environment*. Ellis Horwood Limlted. 192 p.

Vernadsky V.I. (1978). *Living matter*. M.: Science. 358p.

Vetrov V.A., Kuznetsova A.I. (1997). *Microelements in natural media of Lake Baikal region*. RAS, Inst. Geochem. SB RAS Novosibirsk: 234 p.

7

Cu, Pb and Zn Fractionation
in a Savannah Type Grassland Soil

B. Anjan Kumar Prusty*, Rachna Chandra and P. A. Azeez
Environmental Impact Assessment Division
Sálim Ali Centre for Ornithology and Natural History (SACON)
Anaikatty (PO), Coimbatore
India

1. Introduction

Heavy metal contamination in soil has received much attention and extensive research has been conducted on estimation of metals in soils. Several heavy metals are known to accumulate in water, soil, sediments and tissues of organisms (Chaphekar, 1991; Lambou & Williams, 1980; Ramadan, 2003) and cause chronic to acute toxicity in due course of time. For evaluation of the heavy metal burden in the environment, it is not sufficient to measure only total metal concentrations, it is also very important to establish the proportions of heavy metals present in various soil solid phase forms their bioavailability and toxicity. Quantifying the geochemical phases of metals associated with soil is an important step in predicting the ultimate fate, bioavailability, and toxicity of metals (Azeez et al., 2006; Lu et al., 2005; Prusty et al., 1994). Easily soluble fractions that are bioavailable and most mobile in the environment are commonly investigated by single extraction procedures such as that of Deely et al. (1992), while the partitioning of heavy metals between easily and sparingly soluble fractions in soils and sediments is investigated by sequential extraction procedures.

Sequential extractions, although operationally defined, give information about the association of heavy metals with geochemical phases of soil, and hence helps to reveal the distribution of heavy metals in fractions and to assess the mobility and toxicity of metals in soils (Ahnstrom & Parker, 1999; Quevauiller et al., 1993). Several sequential chemical extraction procedures are in practice in speciation studies to assess metals in different environmental matrices (Badri & Aston, 1983; Kersten & Förstner, 1986; Pickering, 1981; Tessier et al., 1979; Young et al., 1992). Element specific methods have also been developed as that of Poulton & Canfield (2005) for iron partitioning. Of these, the five step extraction method by Tessier et al. (1979) is a widely used one, although the disadvantages of this extraction scheme, e.g., non-specificity of extraction (Nirel & Morel, 1990; Reuther, 1999) and re-sorption (Howard & Shu, 1996; Howard & Vandenbrink, 1999) have been well recognized. According to this protocol, metals in soil are fractionated into five geochemical pools, *viz.*, exchangeable (EXC), carbonate- (CA), multiple hydroxide (Fe-Mn oxide), organic matter and sulphide (oxidizable, OM-S), and lithogenic or residual

---

* Corresponding Author

(RES). The mobility and bioavailability of metals decrease approximately in the order of extraction sequence (Prusty et al., 1994) and hence the strength of the chemical reagents used in extraction increases with the sequence. Generally, exchangeable form is considered readily mobile and easily bioavailable, while lithogenic or residual form is considered as incorporated into crystalline lattice of soil minerals and the most inactive. The carbonate, Fe-Mn oxide and organic matter- bound fractions could be relatively active depending on the physical and chemical properties of the medium. Metals bound to sulphides and organic matter are more stable and hard to take part in the geochemical cycle and generally act as a sink and reservoir for pollutants (Prusty et al., 1994; Yuan et al., 2004).

Heavy metal partitioning in soil has become imperative as this acts as a potential environmental indicator and soils enriched in metals may eventually become sources of metal contamination during oxidative weathering and be available for plant uptake. Metal speciation studies have been extensively undertaken in aquatic systems (Fan et al., 2002; Jagadeesh et al., 2006; Li et al., 2000, 2001; Mathew et al., 2003; Prusty et al., 2007a; Shanthi et al., 2003) for assessment of metal mobility and bioavailability. However, less progress has been made in terrestrial environments, mainly owing 1) to the difficulties in measuring metal activities in soil systems, and 2) to the heterogeneous nature of the soil environment, where exposure of organisms to metals occurs through solid, liquid, and gaseous pathways (Nolan et al., 2003). The Keoladeo National Park (KNP), India inspite of being one of the early Ramsar sites in the country is lacking studies on soil characterization and assessment of metal distribution in the soil. The impetus behind the research presented here was to investigate the chemical partitioning of Cu, Pb and Zn in the soil profile of the grassland system in the park. The particular objectives were to 1) examine the distribution of the metals amongst different operationally defined geochemical pools, and 2) assess the variation of metal distribution among years and soil layers.

2. Study area

Field investigations were carried out in the Keoladeo National Park (KNP, Figure 1), Bharatpur, India in summer of 2003, 2004 and 2005. It was declared as a protected area and bird sanctuary in 1956 and later upgraded to a National Park in 1981 (Sharma & Praveen, 2002). This 29 Km^2 park (27°7.6' to 27°12.2'N and 77°29.5' to 77°33.9'E, almost equidistant about 180 km from Delhi and Jaipur), exixting for more than 250 years (Azeez et al., 1992) is one of the early Ramsar sites (Mathur et al., 2005). The park is segmented into 15 blocks or compartments, named alphabetically from A to O, separated by earthen dykes or mud trails, for the ease of management and tourism. The large number of migratory water fowls, the range of habitats (Prusty et al., 2006) clearly distinguished by vegetation types (Davis & van der Valk, 1988) and hydrological parameters (Azeez et al., 2007) available in the area are the distinctive features of the park. About 8.5 km^2 large central depression of the park is (Figure 1) wetland, while the rest is covered by grassland and woodland (Azeez et al., 2000). The grassland is comprised of *Vetiveria zizanioides*, *Desmostachya bipinnata* and *Cynodon dactylon* as characteristic vegetation (Sharma & Praveen, 2002). While the study was conducted in all the habitats, for the scope of this chapter only the results from the grassland habitat were synthesized and presented here.

The area is a part of the Aravalli Supergroup and falls under semi-arid hot dry zone of India (Pal et al., 2000) and experiences four distinct seasons; summer / pre-monsoon (April to June), rainy / monsoon (July to Mid-September), post monsoon (mid-September to mid-November) and winter (mid-November to March, Prusty & Azeez, 2004). Usually the temperature in the Park varies from 1°C to 49°C, showing strong diurnal and seasonal fluctuations. The soil type of the park is clay loam (terrestrial areas) to clayey (aquatic areas) with scattered saline patches in the terrestrial areas.

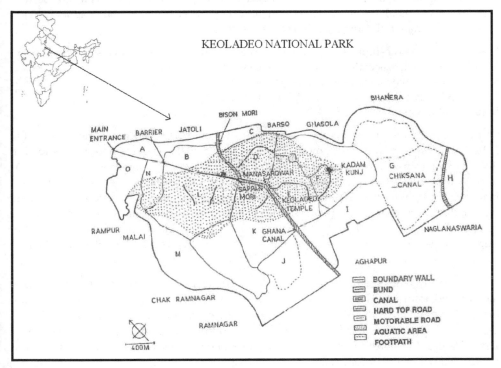

Fig. 1. Study Area Map

3. Methods

3.1 Soil sampling, processing and analyses

Of the total grassland area, three blocks were considered for the present study and the samples were collected in mid June of the year. A trench of 1 m³ was dug up in three locations in each block and the uppermost litter layer was removed by scrapping with a plastic scrapper. Subsequent soil layers from 0, 25, 50, 75 and 100 cm depths were removed using a plastic scoop. The soil samples were packed in pre-cleaned, acid treated and airtight plastic bags and transferred to the laboratory for further processing and analyses. In the laboratory, the soil samples were air-dried at room temperature (Jackson, 1958), homogenized / gently crushed using an agate mortar and pestle and sieved through a standard sieve of 2 mm mesh size (Tandon, 2001). The soil samples with particle size of <2 mm were stored in acid washed plastic containers. Equal proportion of samples from all the three points from each block were thoroughly mixed to get a representative sample per block.

Total Organic Carbon (TOC, %) in soil was estimated following the wet digestion method of Walkley & Black (1934) and the values were converted to the Total Organic Matter (TOM) empirically assuming that TOC comprises 58% of TOM. The Carbonate Carbon (CO_3-C, %) content in soil was estimated following the rapid titration method of Allen (1989). The chemical fractionation of metals was assessed following Sequential Extraction Procedure based on Tessier et al. (1979). It was carried out progressively on an initial weight of 1.0 g of homogenized material using the following extractions:

Step 1. 0.5 M Magnesium chloride adjusted to pH 7.0 with 10% ammonia solution.
Step 2. 1 M Sodium acetate adjusted to pH 5.0 with Acetic acid.
Step 3. 0.04 M Hydroxylamine hydrochloride in 25% Acetic acid.
Step 4. 30% Hydogen peroxide in 0.02 M Nitric acid.
Step 5. Aquaregia digestion.

Of the five steps mentioned above, step 1 to step 4 were performed following Tessier et al. (1979) and step 5 was performed following Ure (1990). The method is intended to distinguish five fractions representing the following phases; Exchangeable (EXC – step 1), Carbonate and specifically adsorbed (CA – step 2), Fe-Mn oxide (Fe-Mn – step 3), Organic matter and sulphide (OM-S – step 4) and Residual / lithogenic (Res – step 5). Aquaregia digestion was performed to estimate the total metal (pseudototal level) content which was compared with the sum of all the five fractions. The geochemical phases at each extraction step are largely operationally defined and indicate relative rather than absolute chemical speciation. The main interpretations are based on the solubility of metals.

Simultaneously blanks and internal standards were also run to verify the precision of the method and accuracy. The precision and bias was generally <10%. The estimated detection limits of the metals in the soil (μg/g) and the recovery rate for all the metals are given in Table 1. Analysis of standardised soil samples showed an average recovery rate of 94.6 ± 1.9%. Values for constituents lower than the method detection limits (< DL) were substituted with DL/2 prior to statistical analysis (Farnham et al., 1998; Ryu et al., 2006). For quality assurance throughout the experiments and analyses, all extracting reagents were prepared using metal free, AnalaR grade chemicals procured from Qualigens Fine Chemicals Division of GlaxoSmithKline Pharmaceuticals Limited, Mumbai and double distilled water prepared using quartz double distillation assembly was used for the reagent preparations. Room temperature was 30°C, while extractions were carried out. Polypropylene centrifuge tubes and bottles were subjected to cleaning procedures prescribed by Laxen & Harrison (1981). The metals in the extracts were analyzed using an AAS (Perkin Elmer AAnalyst 800).

Metal	Cu	Pb	Zn
Detection level in soil (μg/g)	0.01	0.01	0.1
Recovery rate (%)	94.3	92.9	96.7

Table 1. Detection limits for metals

3.2 Statistical analyses

Basic descriptive statistics and two-tail Correlation matrix were performed on the analytical data using "MEGASTAT" to infer the range, distribution and association of different metal fractions among themselves and with TOM and CO_3-C. Univariate tests were performed following the General Linear Model (GLM) to assess variations of the distribution of the metals among fractions, soil layers and years. Post-hoc analysis, i.e. One Way Analysis of Variance (ANOVA) was performed coupled with a test of Least Significant Difference (LSD), only in the cases with significant differences. The statistical tests were performed using SPSS 11.0 (Norušis, 1986).

4. Results and discussions

4.1 Distribution of heavy metals in soil profile

Of the metals studied, Zn was seen highest in soil while Pb was seen the least. The total concentration of Cu in soil ranged from 23.6mg/Kg to 46.4 mg/Kg. The range for Zn was 35.9 mg/Kg to 63.2 mg/Kg and Pb was 8.7 mg/Kg to 13.1 mg/Kg. Cu was seen least at a depth of 25 cm during 2003 and was seen highest in the surface layer during 2004. Pb was found minimum at the depth of 100 cm duing 2005, while maximum was seen at a depth of 50 cm during 2005. The lowest concentration of Zn was at a depth of 25 cm during 2003 and the highest at the bottom of the profile. Two-Way ANOVA test results show that the variation of Cu and Zn was highly significant among the years (Table 2). Of the three metals studied, Cu and Zn had an increasing pattern along the soil profile during all the years (Figure 2). However, a sustained pattern of increase throughout the depth profile was observed duirng 2005 only. Irrespective of the year and metal, the metal level decreased from surface layer to the immediate next layer, i.e. to a depth of 25 cm and then gradually increased till the bottom of the soil profile. The depthwise trend of increase in metals concentration could be associated with their lithogenic nature. The extent of their lithogenic nature is discussed in the subsequent paragraphs. The higher metal levels in the surface layers compared to the subsequent layer might be because of the two reasons, of which the first being the contribution from the plant litters during the process of decomposition. The results of the earlier works by the authors in the KNP reports several species of plants showing notable uptake rates for elements such as Cu, Pb and Zn from the soil (Azeez et al., 2007; Prusty et al., 2007b). The second reason may be the automobile exhausts, they being one of the substantial sources for metals such as Pb (Baker, 1990). KNP has one of the major and busy National High-way (Delhi-Jaipur) traversing beside its immediate boundary. Moreover, its proximity to Bharatpur city might add substantially to the cause.

Metal	Source of variation	F actual	F critical	P value
Cu	Years	6.169	4.459	*0.024
	Soil Layers	0.228	3.838	0.915
Pb	Years	0.666	4.459	0.54
	Soil Layers	2.139	3.838	0.167
Zn	Years	20.432	4.459	*7.18-04
	Soil Layers	1.253	3.838	0.363
* Significant at p<0.05				

Table 2. ANOVA of total metal content in soil

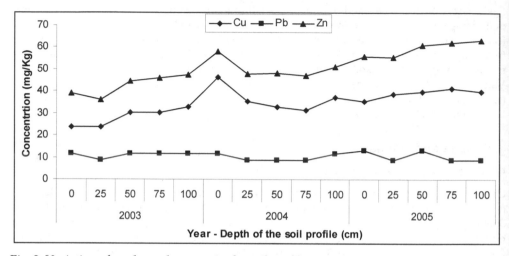

Fig. 2. Variation of total metal content in the soil profile

4.2 Fractionation of heavy metals in soil profile

The association of the three metals to various geochemical phases is presented in Figure 3 to Figure 5. Of the three metals studied, Cu was mainly associated with RES phase while it showed least preferences to the EXC phase (Figure 3). Cu was seen in the range of 30.4% to 76.2% in the residual phase and 0.01% to 0.7% in the EXC phase and the order of the fractions in terms of Cu concentration was Res > Fe-Mn > OM-S > CA > EXC. In contrast, Pb was mostly contained in the multiple hydroxides fraction and least attached with the EXC phase (Figure 4). Pb was seen in the range of 36.6% to 74.0% in the Fe-Mn oxide phase and 0.02% to 7.5% in the EXC phase and the order of fraction in terms of Pb concentration was Fe-Mn > Res > OM-S > CA > EXC. Zn, similar to Cu, had preference towards RES phase and least preference towards EXC phase (Figure 5). Zn was seen in the range of 60.7% to 88.9% in the RES phase, while 0.1% to 1.6% was seen in the EXC phase. The fractions in terms of Zn levels was in the order Res > Fe-Mn > OM-S > CA > EXC. The results indicate two major points: (1) RES phase is the major binding site for Cu and Zn, indicating that the major proportion of the metal is incorporated in the silicate mineral matrix. This may indicate that this element was derived from natural geological sources. (2) Fe-Mn phase is the important binding site for Pb and the reducible Fe and Mn plays a major role in binding these metals. Fe-Mn phase represents the second most significant sink for Cu and Zn after RES phase. It has been shown that in Saline - alkaline soils, as in the case of KNP, the second extraction step (CA phase) may not be effective in removing all the carbonate minerals into solution. Metals extracted in step 3 (Fe-Mn phase) therefore may contain a proportion of the carbonate forms in addition to those bound to Fe-Mn oxides (Maskall & Thornton 1998). The observed higher attachment of Cu and Zn to the RES phase is consistent with observations of Li et al. (2001). The highest affinity of Zn towards RES phase was also reported by Svete et al. (2001) in a study on the chemical partitioning of Zn from a mine area.

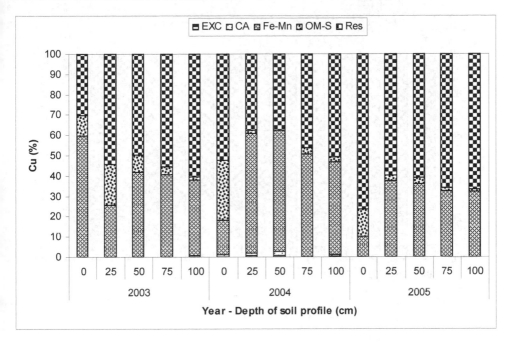

Fig. 3. Cu fractionation in grassland

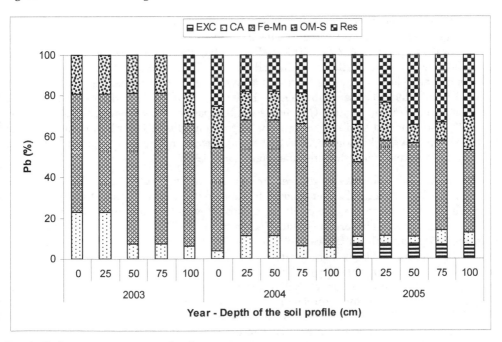

Fig. 4. Pb fractionation in grassland

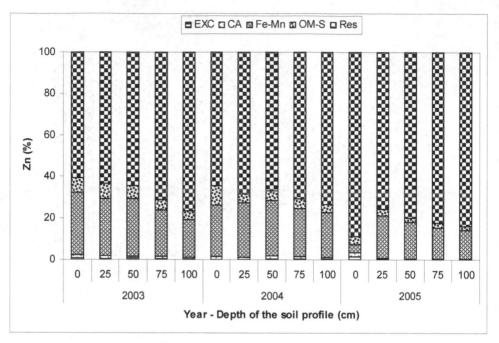

Fig. 5. Zn fractionation in grassland

Although all the metals showed least preference to the exchangeable pool, the order of the metals with average percentage bound to the EXC pool in parenthesis was Cu (0.1 ± 0.2%) < Zn (0.4 ± 0.3%) < Pb (2.4±3.4%). This indicates that the anthropogenic input of Pb is comparatively higher than those of Cu and Zn. As the sampling was done in summer months that have less organic matter in soil as evidenced by our earlier reports, the metals mostly remain attached either with the RES phase and/or with the Fe-Mn oxides phase, and the attachment to organic matter (OM-S) phase is comparatively less. Fan et al. (2002) & Li et al. (2000) also reported similar findings. The apparently greater contribution made by the hydroxylamine hydrochloride-extractable (Fe-Mn) fraction to the Pb compared to Cu and Zn was also reported from other environments (Jones, 1987).

The extent of association among different fractions, and with TOM (%) and CO_3-C (%) is shown in Table 3. Among the non residual fractions only Fe-Mn was negatively correlated with the RES fraction in the case of all the metals. However, in specific cases such as Zn, OM-S was also negatively correlated with RES fraction. The OM-S was positively correlated with the TOM (%) only in the case of Cu (r = 0.673, P<0.05). Similar results were also reported by the authors in their study on heavy metal fractionation in woodland soil (Prusty et al., 2009). However, the negative correlation of CA with CO_3-C (%) needs further investigation. The contribution of RES fraction to the total metal content in soil was supported by their poisitive correlation, which was significant only in the case of Zn (r = 0.743, P<0.05) indicating the contribution of the lithogenic fraction to the total concentration. The proportion of the metals in the RES phase increased and that in the Fe-Mn oxide fraction decreased along the soil profile in case of all the metals.

Cu

	EXC	CA	Fe-Mn	OM-S	Res	SP-PT	TOM (%)	CO$_3$-C (%)
EXC	1.000							
CA	.787*	1.000						
Fe-Mn	.570*	.366	1.000					
OM-S	-.344	-.055	-.592*	1.000				
Res	-.509	-.466	-.826*	.038	1.000			
Total metal	.021	.166	-.396	.018	.455	1.000		
TOM (%)	-.275	-.069	-.617*	.673*	.289	.176	1.000	
CO$_3$-C (%)	-.393	-.110	-.107	-.250	.312	.157	-.070	1.000

Pb

	EXC	CA	Fe-Mn	OM-S	Res	SP - PT	TOM (%)	CO$_3$-C (%)
EXC	1.000							
CA	-.435	1.000						
Fe-Mn	-.773*	.281	1.000					
OM-S	-.423	.126	.204	1.000				
Res	.744*	-.659*	-.847*	-.461	1.000			
Total metal	.017	-.247	.037	.218	.012	1.000		
TOM (%)	.228	-.070	-.445	.219	.276	.490	1.000	
CO$_3$-C (%)	.446	-.160	-.203	-.780*	.396	-.359	-.070	1.000

Zn

	EXC	CA	Fe-Mn	OM-S	Res	SP - PT	TOM (%)	CO$_3$-C (%)
EXC	1.000							
CA	.593*	1.000						
Fe-Mn	-.395	.238	1.000					
OM-S	.098	.650*	.612*	1.000				
Res	.218	-.441	-.964*	-.791*	1.000			
Total metal	-.216	-.538*	-.683*	-.583*	.743*	1.000		
TOM (%)	.664*	.594*	-.408	.352	.180	.160	1.000	
CO$_3$-C (%)	-.129	-.250	-.326	-.382	.381	.362	-.070	1.000

45 Sample size, ± .514 critical value, $P < 0.05$ (two-tail), * Significant

EXC = Exchangeable metal, CA = Carbonate bound metal, Fe-Mn = Iron – Manganese bound metal, OM-S = Bound to organic matter and sulphur, Total metal = Pseudototal metal, TOM = Total Organic Matter, CaCO$_3$ = Carbonate carbon

Table 3. Correlation matrix of metal fractions in the soil

The Univariate test performed following the GLM showed that the distribution of metals was significant only among fractions ($P<0.05$) and hence the Post-hoc analysis (One-Way ANOVA - LSD) was performed on the fractions and the results are shown in Table 4 to Table 6. Fe-Mn phase was found to contribute significantly to the reported variability in the case of all the metals studied. However, there are other additional phases in case of specific metals, for e.g., in the case of Cu, RES was also a significant contributor to the cause (Table 4). In the case of Pb, it was EXC and CA (Table 5), and in the case of Zn, the significant contributor was OM-S and RES (Table 6). As indicated earlier, the findings in the case of Cu and Zn indicate the lithogenic nature of the metals, while in the case of Pb it indicates to the anthorpogenic influences. The significant contribution of CA in the case of Pb may be due to the relatively high pH values in soil at the study site, which have been elevated due to the release of Ca and Carbonate compounds from the soil. Previous study undertaken by the authors report saline and alkaline patches throughout the terrestrial area in the park wherein, the CO_3-C content in grassland soil in KNP ranged from 5.4% to 27.5%. The highest pH in the grassland system in KNP was reported to be 9.97. Maskall & Thornton (1998) reported similar observations and ascribe these to the release of Ca from soil. It can be explained by specific adsorption, which is more important for Pb than other two metals (Borůvka et al., 1997).

Multiple Comparisons

Dependent Variable: CU

LSD

(I) FRACTION	(J) FRACTION	Mean Difference (I-J)	Std. Error	Sig.	95% Confidence Interval	
					Lower Bound	Upper Bound
EXC	CA	-.2417	3.42211	.944	-7.0669	6.5835
	Fe-Mn	-38.8850*	3.42211	.000	-45.7102	-32.0598
	OM-S	-6.7867	3.42211	.051	-13.6119	.0385
	RES	-53.4141*	3.42211	.000	-60.2393	-46.5889
CA	EXC	.2417	3.42211	.944	-6.5835	7.0669
	Fe-Mn	-38.6433*	3.42211	.000	-45.4685	-31.8181
	OM-S	-6.5450	3.42211	.060	-13.3702	.2802
	RES	-53.1724*	3.42211	.000	-59.9976	-46.3472
Fe-Mn	EXC	38.8850*	3.42211	.000	32.0598	45.7102
	CA	38.6433*	3.42211	.000	31.8181	45.4685
	OM-S	32.0983*	3.42211	.000	25.2731	38.9235
	RES	-14.5291*	3.42211	.000	-21.3543	-7.7039
OM-S	EXC	6.7867	3.42211	.051	-.0385	13.6119
	CA	6.5450	3.42211	.060	-.2802	13.3702
	Fe-Mn	-32.0983*	3.42211	.000	-38.9235	-25.2731
	RES	-46.6274*	3.42211	.000	-53.4526	-39.8022
RES	EXC	53.4141*	3.42211	.000	46.5889	60.2393
	CA	53.1724*	3.42211	.000	46.3472	59.9976
	Fe-Mn	14.5291*	3.42211	.000	7.7039	21.3543
	OM-S	46.6274*	3.42211	.000	39.8022	53.4526

*. The mean difference is significant at the .05 level.

Table 4. LSD of Cu – fractions

Multiple Comparisons

Dependent Variable: PB
LSD

(I) FRACTION	(J) FRACTION	Mean Difference (I-J)	Std. Error	Sig.	95% Confidence Interval Lower Bound	Upper Bound
EXC	CA	-6.2347*	3.06070	.045	-12.3390	-.1303
	Fe-Mn	-51.8199*	3.06070	.000	-57.9242	-45.7155
	OM-S	-14.3561*	3.06070	.000	-20.4604	-8.2517
	RES	-15.6058*	3.06070	.000	-21.7102	-9.5014
CA	EXC	6.2347*	3.06070	.045	.1303	12.3390
	Fe-Mn	-45.5852*	3.06070	.000	-51.6896	-39.4808
	OM-S	-8.1214*	3.06070	.010	-14.2258	-2.0170
	RES	-9.3711*	3.06070	.003	-15.4755	-3.2668
Fe-Mn	EXC	51.8199*	3.06070	.000	45.7155	57.9242
	CA	45.5852*	3.06070	.000	39.4808	51.6896
	OM-S	37.4638*	3.06070	.000	31.3594	43.5682
	RES	36.2141*	3.06070	.000	30.1097	42.3184
OM-S	EXC	14.3561*	3.06070	.000	8.2517	20.4604
	CA	8.1214*	3.06070	.010	2.0170	14.2258
	Fe-Mn	-37.4638*	3.06070	.000	-43.5682	-31.3594
	RES	-1.2497	3.06070	.684	-7.3541	4.8546
RES	EXC	15.6058*	3.06070	.000	9.5014	21.7102
	CA	9.3711*	3.06070	.003	3.2668	15.4755
	Fe-Mn	-36.2141*	3.06070	.000	-42.3184	-30.1097
	OM-S	1.2497	3.06070	.684	-4.8546	7.3541

*. The mean difference is significant at the .05 level.

Table 5. LSD of Pb – Fractions

Multiple Comparisons

Dependent Variable: ZN
LSD

(I) FRACTION	(J) FRACTION	Mean Difference (I-J)	Std. Error	Sig.	95% Confidence Interval Lower Bound	Upper Bound
EXC	CA	-.5969	1.80145	.741	-4.1898	2.9959
	Fe-Mn	-20.8475*	1.80145	.000	-24.4403	-17.2546
	OM-S	-4.2336*	1.80145	.022	-7.8265	-.6407
	RES	72.2304*	1.80145	.000	-75.8233	-68.6375
CA	EXC	.5969	1.80145	.741	-2.9959	4.1898
	Fe-Mn	-20.2505*	1.80145	.000	-23.8434	-16.6577
	OM-S	-3.6367*	1.80145	.047	-7.2295	-.0438
	RES	-71.6335*	1.80145	.000	-75.2263	-68.0406
Fe-Mn	EXC	20.8475*	1.80145	.000	17.2546	24.4403
	CA	20.2505*	1.80145	.000	16.6577	23.8434
	OM-S	16.6139*	1.80145	.000	13.0210	20.2067
	RES	-51.3829*	1.80145	.000	-54.9758	-47.7901
OM-S	EXC	4.2336*	1.80145	.022	.6407	7.8265
	CA	3.6367*	1.80145	.047	.0438	7.2295
	Fe-Mn	-16.6139*	1.80145	.000	-20.2067	-13.0210
	RES	-67.9968*	1.80145	.000	-71.5897	-64.4039
RES	EXC	72.2304*	1.80145	.000	68.6375	75.8233
	CA	71.6335*	1.80145	.000	68.0406	75.2263
	Fe-Mn	51.3829*	1.80145	.000	47.7901	54.9758
	OM-S	67.9968*	1.80145	.000	64.4039	71.5897

*. The mean difference is significant at the .05 level.

Table 6. LSD of Zn – Fractions

The heavy metal partitioning in soil could be used to determine its mobility and possible sources. Fractionation of total metal contents might give indications about the origin of the metals. The levels in the EXC, CA and Fe-Mn fraction may indicate pollution from anthropogenic origin, and those in OM-S and RES fraction are relatively immobile. Hence, the Fe-Mn phase being the common factor in all the metals, needs further attention from the point of view of metal mobility and bioavailability. Due to their high scavenging capacity, Fe and Mn oxides has been recorded as significant heavy metal sink in soil. Although the dynamics of metal scavenging by Fe and Mn oxides is still poorly understood, assuming that the extraction using Hydroxylamine hydrochloride with acetic acid is an appropriate indicator of metals associated with amorphous iron and manganese oxides, this process seems to exert a significant control on the metals studied in this grassland system. Although the attachment of metals to Fe-Mn oxides indicates that metals are relatively immobilized, slight chemical changes in the ambient conditions could result in their likely mobility and easy absorption by plants. pH and redox changes are the two crucial factors in this regard.

5. Conclusions

The analysis of the extracts produced by a sequential extraction procedure allowed the determination of metals in different fractions of the soil samples. Of all the metals studied the proportion of Pb was maximum in EXC phase and the order of metals in this fraction was Cu < Zn < Pb indicating their relative anthropogenic input. Fe-Mn oxide phase was found to be the common sink for all the metals under study and a significant contributor to the reported variability of each of these metals. However, to confirm this more detailed investigations of the mechanisms controlling the distribution and mobility of different metal species are required.

6. Acknowledgenents

We acknowledge the Council of Scientific and Industrial Research (CSIR), India for the financial support as Senior Research Fellowship (Grant No.- 9/845 (4)/06 – EMR-I) to the first author. M/s Brijendra Singh, Randhir Singh and Rajesh Singh assisted during the trench (soil) sampling. Ms. Jayalakshmi was helpful during the sequential chemical extraction of the soil samples.

7. References

Ahnstrom, Z. S., & Parker, D. R. (1999). Development and assessment of a sequential extraction procedure for the fractionation of soil cadmium. *Soil Science Society of America Journal*, Vol. 63, pp. 1650-1658.
Allen, S. E. (1989). *Chemical Analysis of Ecological Materials* (2nd Edition), Blackwell Scientific Publications, London, pp. 368.
Azeez, P. A., Nadarajan, N. R., & Mittal, D. D. (2000). The impact of a monsoonal wetland on ground water chemistry. *Pollution Research*, Vol. 19, No. 2, pp. 249-255.
Azeez, P. A., Nadarajan, N. R., & Prusty, B. A. K. (2007). Macrophyte decomposition and changes in water quality. In: *Environmental Degradation and Protection*,

Eds: K. K. Singh, A. Juwarkar, A. K. Singh, pp. 115-156, MD Publications, New Delhi.

Azeez, P. A., Prusty, B. A. K., & Jagadeesh, E. P. (2006). Chemical speciation of metals in environment, its relevancy to ecotoxicological studies and the need for biosensor development. *Journal of Food, Agriculture and Environment*, Vol. 4, No. 3 & 4, pp. 235-239.

Azeez, P. A., Ramachandran, N. K., & Vijayan, V. S. (1992). The socioeconomics of the villagers around Keoladeo National Park, Bharatpur, Rajasthan. *International Journal of Ecology and Environmental Sciences*, Vol. 18, pp. 169-179.

Badri, M. A., & Aston, S.R. (1983). Observations on heavy metal geochemical associations in polluted and non-polluted estuarine sediments. *Environmental Pollution (Series. B)*, Vol. 6, pp. 181-193.

Baker, D. E. (1990). Copper. In: *Heavy metals in soils*, Ed. B. J. Alloway, pp. 151-176, Blackie & Sons, London.

Borůvka, L., Krištoufká, S., Kozák, J., & Huan-Wei, Ch. (1997). Speciation of Cadmium, lead and xinc in heavily polluted soils. *Rostlina Vyroba*, Vol. 43, No. 4, pp. 187-192.

Chaphekar, S. B. (1991). An overview on bioindicators. *Journal of Environmental Biology*, Vol. 12, pp. 163-168.

Davis, C. B., & van der Valk, A. (1988). *Ecology of a semitropical monsoonal wetland in India: The Keoladeo National Park, Bharatpur, Rajasthan*. Final Report, September 1988, Ohio State University, 104 pp.

Deely, J. M., Tunnicliss, J. C., Orange, C. J., & Edgerley, W. H. L. (1992). Heavy metals in surface sediments of Waiwhetu stream, Lower Hutt, New Zealand. *New Zealand Journal of Marine and Freshwater Research*, Vol. 26, pp. 417-427.

Fan, W., Wang, W. X., Chen, J., Li, X. D., & Yen, Y. F. (2002). Cu, Ni and Pb speciation in surface sediments from a contaminated bay of northern China. *Baseline Marine Pollution Bulletin*, Vol. 44, pp. 816-826.

Farnham, I., Smiecinski, A., & Singh, A. K. (1998). Handling chemical data below detection limits for multivariate analysis of groundwater. *First International Conference on Remediation of Chlorinated and Recalcitrant Compounds*. Monterey, CA, pp. 99-104.

Howard, J. L., & Shu, J. (1996). Sequential extraction analysis of heavy metals using a chelating agent (NTA) to counteract resorption. *Environmental Pollution*, Vol. 91, No. 1, pp. 89-96.

Howard, J. L., & Vandenbrink, W. J. (1999). Sequential extraction analysis of heavy metals in sediments of variable composition using nitrilotriacetic acid to counteract resorption. *Environmental Pollution*, Vol. 106, pp. 285-292.

Jackson, M. L. (1958). *Soil Chemical Analysis*. Constable & Co Ltd, London, pp. 498.

Jagadeesh, E. P., Azeez, P. A., & Banerjee, D. K. (2006). Modeling chemical speciation of copper in River Yamuna at Delhi, India. *Chemical Speciation and Bioavailbility*, Vol. 18, No. 2, pp. 61-69.

Jones, J. M. (1987). Chemical fractionation of Copper, Lead and Zinc in ombrotrophic peat. *Environmental Pollution*, Vol. 48, pp. 131-144.

Kersten, M., & Förstner, U. (1986). Chemical fractionation of heavy metals in anoxic estuarine and coastal sediments. *Water Science Technology*, Vol. 18, pp. 121-130.

Lambou, V. W., & Williams, L. R. (1980). Biological monitoring of hazardous wastes in aquatic systems. *Second Interagency Workshop on in-situ Water Sensing: Biological Sensors*, Pensacola Beach, Florida, pp. 11-18.

Laxen, D. P. H., & Harrison, R. M. (1981). Cleaning methods for Polythene containers prior to the determination of trace metals in fresh water samples. *Analytical Chemistry*, Vol. 53, pp. 345-350.

Li, X. D., Shen, Z., Wai, O. W. H., & Li, Y-S. (2000). Chemical partitioning of heavy metal contaminants in sediments of the Pearl River estuary. *Chemical Speciation and Bioavailability*, Vol. 12, No. 1, pp. 17-25.

Li, X. D., Shen, Z., Wai, O. W. H., & Li, Y-S. (2001). Chemical forms of Pb, Zn and Cu in the sediment profiles of the Pearl River estuary. *Marine Pollution Bulletin*, Vol. 42, No. 3, pp. 215-223.

Lu, A., Zhang, S., & Shan, X. (2005). Time effect on the fractionation of heavy metals in soils. *Geoderma*, Vol. 125, pp. 225-234.

Maskall, J. E., & Thornton, I. (1998). Chemical partitioning of heavy metals in soils, clays and rocks at historical lead smelting sites. *Water, Air and Soil Pollution*, Vol. 108, pp. 391-409.

Mathew, M., Mohanraj, R., Azeez, P. A., & Pattabhi, S. (2003). Speciation of heavy metals in bed sediments of wetlands in urban Coimbatore, India. *Bulletin of Environmental Contamination and Toxicology*, Vol. 70, pp. 800-808.

Mathur, V. B., Sinha, P. R., & Mishra, M. (2005). *Keoladeo National Park World Heritage Site*. Technical Report No. 5, UNESCO-IUCN-Wild Life Institute of India, Dehradun, India.

Nirel, P. M. V., & Morel, F. M. M. (1990). Pitfalls of sequential extractions. *Water Research*, Vol. 24, pp. 1055-1056.

Nolan, A. L., Lombi, E., & McLaughlin, M. J. (2003). Metal bioaccumulation and toxicity in soils – why bother with speciation? *Austalian Journal of Chemistry*, Vol. 56, pp. 77-91.

Norušis, M.J. (1986). *SPSS/PC+4.0 Base Manual – Statistical Data Analysis*. SPSS Inc.

Pal, D. K., Bhattacharyya, T., Deshpande, S. B., Sarma, V. A. K., & Velayutham, M. (2000). *Significance of minerals in soil environment of India*, NBSS Review Series 1. National Bureau of Soil Survey & Land Use Planning, Nagpur, India, pp. 68.

Pickering, W. F. (1981). Selective chemical extraction of soil components and bound metal species. *CRC Critical Reviewes in Analytical Chemistry*, Vol. 12, pp. 233.

Poulton, S. W., & Canfield, D. E. (2005). Development of a sequential extraction procedure for iron: implications for iron partitioning in continentally derived particulates. *Chemical Geology*, Vol. 214, pp. 209-221.

Prusty, B. A. K., & Azeez, P.A. (2004). Seasonal variation in water chemistry of a monsoonal wetland. *Souvenir of the National Conference on Water Vision-2004*, PSGR Krishnamal College for Women, Coimbatore, 06–08 October, pp. 47

Prusty, B. A. K., Azeez, P. A., & Jagadeesh, E. P. (2007b). Alkali and Transition Metals in Macrophytes of a Wetland System. *Bulletin of Environmental Contamination and Toxicology*, Vol. 78, No. 5, pp. 405-410 (May 2007).

Prusty, B. A. K., Azeez, P. A., Vasanthakumar, R., & Jayalakshmi, V. (2006). Carbon and nitrogen dynamics in the soil system of a wetland-terrestrial ecosystem complex.

Proceedings of National Conference on Environment and Sustainable Development (Ed: R. Mohanraj), Bharathidasan University, Trichirappalli, India, 19-17 February, pp. 12.

Prusty, B. A. K., Jagadeesh, E. P., Azeez, P. A., & Banerjee, D. K. (2007a). Distribution and chemical speciation of select transition metals in the waters of River Yamuna, New Delhi. *Environmental Science: An Indian Journal*, Vol. 2, No. 3, pp. 139-144 (September 2007).

Prusty, B. A. K., Chandra R., & Azeez, P. A. (2009). Distribution and chemical partitioning of Cu, Pb and Zn in the soil profile of a semi arid dry woodland. *Chemical Speciation and Bioavailability*, Vol. 21, No. 3, pp. 141-151 (September 2009).

Prusty, B. G., Sahu, K. C., & Godgul, G. (1994). Metal contamination due to mining and milling activities at the Zawar zinc mine, Rajasthan, India. 1. Contamination of stream sediments. *Chemical Geology*, Vol. 112, pp. 275-291.

Quevauiller, P. H., Rauret, G., & Griepink, B. (1993). Conclusions of the workshop: single and sequential extraction in sediments and soils. *International Journal of Environmental and Analytical Chemistry*, Vol. 57, pp. 135-150.

Ramadan, A. A. (2003). Heavy metal pollution and biomonitoring plants in lake Manzala, Egypt. *Pakistan Journal of Biological Sciences*, Vol. 6, pp. 1108-1117.

Reuther, R. (1999). Trace metal speciation in aquatic sediment: methods, benefits and limitations. In: *Manual of Bioassessment of Aquatic Sediment Quality*, Eds: A. Mudroch, J. M. Azcue and P. Mudroch, Lewis, Boca Raton, pp. 1-54.

Ryu, J. S., Lee, K. S., Kim, J. H., Ahn, K. G., & Chang, H. W. (2006). Geostatistical analysis for hydrochemical characterization of the Han River, Korea: Identification of major factors governing water chemistry. *Bulletin of Environmental Contamination and Toxicology*, Vol. 76, No. 1, pp. 01-07.

Shanthi, K., Ramasamy, K., & Lakshmanaperumalsamy, P. (2003). Sediment quality of Singanallur wetland in Coimbatore, Tamil Nadu, India. *Bulletin of Environmental Contamination and Toxicology*, Vol. 70, pp. 372-378.

Sharma, S., & Praveen, B. (2002). *Management Plan: Keoladeo National Park, Bharatpur*, Plan Period 2002-2006, Department of Forests and Wildlife, Rajasthan, pp. 182.

Svete, P., Milacic, R., & Pihlar, B. (2001). Pertitioning of Zn, Pb and Cd in river sediments from a lead and zinc mining area using the BCR three-step sequential extraction procedure. *Journal of Environmental Monitoring*, Vol. 3, pp. 586-590.

Tandon, H. L. S. (2001). *Methods of analysis of soils, plants, waters and fertilizers*, Fertilizer Development and Consultation Organization (FDCO), New Delhi, pp. 143.

Tessier, A., Campbell, P. G. C., & Bisson, M. (1979). Sequential extraction procedure for the speciation of particulate trace metals. *Analytical Chemistry*, Vol. 51, pp. 844-851.

Ure, A. M. (1990). Methods of analysis for heavy metals in soils. In: *Heavy Metals in Soils*, Ed: B. J. Alloway, Blackie and John Wiley & Sons, New York, pp. 39-80.

Walkley, A., & Black, I.A. (1934). An examination of the Degljareff method for determining soil organic matter and a proposed modification of the chromic acid titration method. *Soil Sci*, Vol. 37, pp. 29-38.

Young, L. B., Dutton, M., & Pick, F. R. (1992). Contrasting two methods for determining trace metal partitioning in oxidized lake sediments. *Biogeochemistry*, Vol. 17, pp. 205-219.

Yuan, C., Jiang, G., Liang, L., Jin, X. and Shi, J. (2004). Sequential extraction of some heavy
    metals in Haihe river sediments, People's' Republic of China. *Bulletin of
    Environmental Contamination and Toxicology*, Vol. 73, pp. 59-66.

8

Evaluating the Effects of Radio-Frequency Treatment on Rock Samples: Implications for Rock Comminution

Arthur James Swart
Vaal University of Technology
Republic of South Africa

1. Introduction

"You find remedy in the thorniest tree". This Arabic proverb well illustrates that scientific solutions to well defined engineering problems are often hard to find, resulting in much frustration and anguish. This has also proved true in the mineral processing industry, where numerous exigent scientific endeavours have sought to improve rock comminution. Comminution may be divided into two steps; the reduction of large materials to a size suitable for grinding (termed crushing) and the reduction of crushed material into powder (termed grinding or milling). Comminution efficiency is currently low and is based on the absolute ratio of energy required to generate new surface area relative to the total mechanical energy input (Tromans, 2008). Current comminution techniques need to be enhanced if a higher efficiency is to be realized.

Mineral liberation efficiency subsequently relates to the amount of energy required to release a certain percentage of valuable minerals from the gangue (waste material) through rock comminution methods. The major source of this energy generation is fossil fuels, coal, natural gas and oil, which are still expected to meet about 84% of energy demand in 2030 (Shafiee & Topal, 2009). However, concerns continue to be raised regarding the burning of fossil fuels as a contributor to rising atmospheric concentrations of carbon dioxide (CO_2), which may contribute to climate change (Wolde-Rufael, 2010). Furthermore, it is estimated that in a mining-intensive country, the minerals processing industry accounts for approximately 18% of the national energy consumption. This process is currently inherently inefficient, with less than 3% of the energy input directly involved in rock breakage and liberation (Moran, 2009). Subsequently, the importance of coal in energy generation and as a possible source of global warming necessitates the use of alternative methods to reduce the amount of energy used by mining industries (increased power efficiency) while at the same time recovering the same (or higher) percentage of valuable minerals (higher throughput) (Wang & Forssberg, 2007). This thorny dilemma continues to frustrate researchers around the globe within the fields of Metallurgical, Mechanical and Electrical Engineering.

Current research studies have found that the mineral liberation process can be enhanced through the use of pulsed power, ultrasound pre-treatment and microwave pre-treatment of run of mine ore (Gaete-Garretón et al., 2000; Wilson et al., 2006; Jones et al., 2007; Wang & Forssberg, 2007). Ore, from the mining operation, goes through a process that separates the valuable minerals from the gangue. This process usually involves crushing, grinding (or milling), separation and extraction where the gangue is usually discarded in tailings piles (Perkins, 1998). These electrical methods, which are used to enhance the mineral liberation process, each have their own advantages and disadvantages, which are discussed in this chapter. However, this research proposes a different technique involving radio-frequency (RF) power, which may have positive implications for rock comminution and mineral liberation.

The main purpose of this research is to evaluate the effect that RF power exerts on rock samples, with particular focus on textural changes. This evaluation aims to determine if RF power weakens mineral grain boundaries, subsequently leading to improved rock comminution and mineral liberation. This may result in significant reductions of energy consumption of current comminution and mineral liberation equipment. This chapter will firstly define rock comminution and mineral liberation. Rocks used in this research are then presented along with current electrical treatment techniques which are applied to enhance rock comminution. A proposed new technique is substantiated with the practical setup of the equipment being introduced. Results of treating specific rock samples with RF power are presented quantitatively (electrical properties, surface temperature rise, particle screen analysis and SEM photomicrographs).

2. Rock comminution and mineral liberation

The processes used for the purification and enhancement of an ore, to satisfy the needs of downstream applications, are collectively referred to as mineral beneficiation (De Waal, 2007). The aim of mineral beneficiation is to separate ore minerals from gangue minerals, producing as pure as possible a concentrate of ore minerals distinct from tailings. The process is undertaken in steps where the run of mine ore first undergoes comminution followed by separation/concentration.

2.1 Comminution

The aim of comminution is to liberate the ore minerals from the gangue by breaking the rock up into smaller particles until there are loose particles of ore mineral (Wills, 1992; Sadrai et al., 2006). Comminution is essentially the size reduction of the fragments of rock/ore (Wills, 1992). Comminution is effected by compression, impact and abrasion, through crushing or grinding/milling. The process usually involves several steps, each comprising a small reduction ratio of three to six. An example of a mineral processing line is demonstrated in Fig. 1. In this example, a vibrating feeder serves the purpose of making coarse separations of mining ores (200 – 800 mm in diameter) and providing a consistent, even supply of rock material to the jaw crusher. The jaw crusher breaks this material down to a particle size of approximately 10 – 100 mm. The next stage, the ball mill, is used for grinding various ore and other materials down to particle sizes of around 100 µm. The stages which follow (classifier to rotary dryer) are used to separate the valuable minerals from the gangue, and are discussed under section 2.2. The physical properties of minerals that have the most influence on the comminution process are hardness, tenacity, cleavage, fracture and common form.

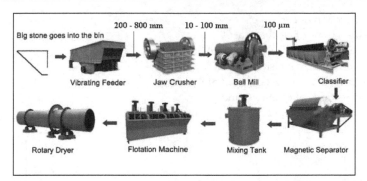

Fig. 1. Mineral processing line (Henan Chuangxin Building-material Equipment Co, 2009)

Hardness is defined as the degree of resistance of a given mineral to scratching, indicating the strength of the bonds that hold the mineral's atoms together (Skinner & Porter, 1992; Chernicoff & Fox, 1997; Klein, 2002; Thompson & Turk, 2007). The hardness of a mineral is tested by scratching the unknown mineral with a series of minerals or substances with known hardness and is one of the most useful diagnostic properties of minerals (Tarbuck & Lutgens, 1999).

Tenacity is a mineral's physical reaction to stress such as crushing, bending, breaking, or tearing (Klein, 2002). Certain minerals react differently to each type of stress. Since tenacity is composed of several reactions to various stresses, it is possible for a mineral to have more than one form of tenacity. The different forms of tenacity are (Rapp, 2009):

- Brittle - If a mineral is hammered and the result is a powder or small crumbs, it is considered brittle. Brittle minerals leave a fine powder if scratched, which is the way to test a mineral to see if it is brittle. Majority of all minerals are brittle. Minerals that are not brittle may be referred to as non-brittle minerals.
- Sectile - Sectile minerals can be separated with a knife into thin slices, much like wax (e.g. gold).
- Malleable - If a mineral can be flattened out into thin sheets by pounding it with a hammer, it is malleable. All true metals are malleable (e.g. gold).
- Ductile - A mineral that can be stretched into a wire is ductile. All true metals are ductile.
- Flexible but inelastic - Any minerals that can be bent, but remains in the new position after it is bent are flexible but inelastic. If the term flexible is singularly used, it implies flexible but inelastic (e.g. chlorite).
- Flexible and elastic - When flexible and elastic minerals are bent, they spring back to their original position. All fibrous minerals and some acicular and flaky minerals belong in this category (e.g. mica).

Cleavages occur when some crystals break in one or more smooth plane surfaces, whose orientation is determined by the regular atomic structure of the crystal (Klein, 2002; Wenk & Bulakh, 2004). Certain minerals fracture with an uneven surface when broken, while others split or cleave along distinctive crystallographic planes. Cleavage is thus the ability of a mineral to break, when struck, along preferred directions (Skinner & Porter, 1992; McGeary et al., 2001). Fig. 2 illustrates seven possible types of mineral cleavage which are:

- A – One direction of cleavage;
- B – Two directions of cleavage at 90°;
- C – Two directions of cleavage not at 90°;
- D – Three directions of cleavage at 90°;
- E – Three directions of cleavage not at 90°;
- F – Four directions of cleavage; and
- G – Six directions of cleavage.

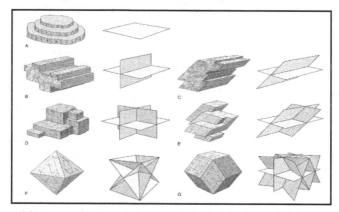

Fig. 2. Seven possible types of mineral cleavage (Wenk & Bulakh, 2004)

Cleavage is tested by striking or hammering a mineral, and is classified by the number of surfaces it produces and the angles between adjacent surfaces (Chernicoff & Fox, 1997). A mineral tends to break along certain planes because the bonding between atoms is weaker there. For example, quartz has equally strong bonds in all directions and would thus have no cleavage, whereas micas are easily split apart into sheets due to the fact that the bonding between adjacent atomic sheets is weak. Cleavage is one of the most useful diagnostic tools because it is identical for a given mineral from one sample to another. It is especially useful for identifying minerals when they appear as small grains in rocks (McGeary et al., 2001).

Some minerals have poorly defined cleavages, while others may not even show any at all. When broken, these minerals cause fractures in that they break on generally irregularly oriented curved surfaces, decided more by stress distribution in the crystal at the time of rupture than by the atomic structure of the mineral (Klein, 2002; Wenk & Bulakh, 2004). Fracture is thus the way a substance breaks when not controlled by cleavage and is the most common type of fracture for minerals (McGeary et al., 2001; Thompson & Turk, 2007). Fracture may appear as a jagged, irregular or rough surface or as a curved, shell-shaped (conchoidal) surface (Chernicoff & Fox, 1997).

Minerals may further be identified by their **common form**. The term form is often used to indicate general outward appearance (Klein, 2002). In crystallography, external shape is denoted by the word habit, whereas the term form is used in a special and restricted sense. Thus a form consists of a group of crystal faces, all of which have the same relation to the elements of symmetry and display the same chemical and physical properties. The term common form is used synonymously to the term habit and refers to the external shape in which a mineral commonly occurs.

2.2 Separation

Automated mineral beneficiation, which essentially involves the separation of specific desired minerals from a crushed/milled mixture of minerals, usually exploits properties such as differences in densities, magnetic susceptibility, electrical conductivity, surface reactivity, refractive index, and fluorescence (Wills, 1992; De Waal, 2007). The most common processes include (Wills, 1992):

- Gravity separation – exploiting the density differences between minerals, and their response to gravity and resistance to motion in a fluid such as water. Typical apparatus includes jigs, Humphries spirals, Reichert cones, sluices, and shaking tables.
- Dense medium separation – exploiting density differences where minerals are introduced to a dense liquid or suspension, in which some minerals will float and others sink, thus effecting separation. A wide variety of separation vessels are employed in industry including some that incorporate a centrifugal aspect to expedite the process.
- Froth flotation – exploiting differences in the surface properties of different types of minerals. Here minerals are exposed to a solution which renders some of the minerals hydrophobic and other hydrophilic. Air is bubbled through the solution in which the minerals are suspended, resulting in separation because the hydrophilic ones settle to the bottom of the solution whereas the hydrophobic minerals can be skimmed off with the soapy froth at the surface.
- Magnetic separation – exploiting the differences in magnetic susceptibility of minerals through use of strong magnetic forces that can be adjusted to separate minerals of differing susceptibility.
- Electrostatic/high tension separation – exploiting differences in electrical conductivity of minerals, in which a charge builds up in non-conductive minerals causing them to stick to charged surfaces, whereas conductive particles do not stick to such surfaces.

Before any of the above separation process can be effective, proper liberation of the ore minerals is essential. Complete liberation is seldom achieved in practice (Wills, 1992; King, 2001), which implies that most particles will comprise both ore and gangue material, with their response to any separation technique being uncertain. The degree of liberation can be described using the following terms:

- A completely liberated particle is one that consists of only one type of mineral, either ore mineral or gangue mineral.
- A middling is a particle that consists of two or more different types of minerals, i.e. it is incompletely liberated.
- Middlings can be further classified into attached mineral (binary, ternary, etc) or enclosed minerals.
- The degree of liberation can also described in terms of what is called particle grade. For example, a liberated particle comprising 100% ore mineral will have a particle grade of 100%, whereas a middling particle consisting of 25% ore mineral and 75% gangue mineral will have a particle grade of 25%.

3. Rocks used in this research

Rocks are composed of minerals (Chernicoff & Fox, 1997) and are called monomineralic when they contain only one mineral (Best & Christiansen, 2001). Ores are essentially rocks that contain one or more type of mineral coveted for its metal content or its physical properties for industrial use (Wills, 1992). The coveted minerals in the ores are called ore minerals (if they contain useful metals) or industrial minerals (if they have useful physical properties). Woollacot and Eric (1994) classify mined material into three categories:

- Mined material consisting of useful rock or soil, where the rock/soil has value in its natural form, e.g. as aggregate or filler material.
- Mined material containing industrial minerals, where the value lies in one or more minerals within the rock that must be liberated and separated from the rock, e.g. diamond in kimberlite, crysotile in greenstones, wollastonite in skarn, etc.
- Mined material containing value-bearing minerals, where the value lies in constituents of one or more minerals within the rock (ore) and the constituent (metal) needs to be extracted from the mineral after the latter has been liberated and separated from the rock (ore), e.g. extraction of copper from copper-bearing minerals such as chalcopyrite ($CuFeS_2$) and bornite (Cu_5FeS_4) occurring as minerals in copper ore.

Rocks may be classified into three groups (see Fig. 3) based on their mode of formation. These are:

- **Igneous rocks:** formed by the solidification/crystallization of mainly molten silicate material called magma or lava (Chernicoff & Fox, 1997; Walther, 2005). These rocks consist of tightly interlocked crystals, where the size of the crystals range from <0.06 mm (as in the case of those crystallized from lava at the surface of the earth) to ±10 mm (as in the case of those crystallized from slow cooling magma deep in the earth's crust). In addition, there are very coarse-grained igneous rocks (pegmatites) which crystallized from magma containing high proportions of volatile material.
- **Sedimentary rocks:** formed by the solidification of loose material on the earth's surface (Skinner & Porter, 1992; Chernicoff & Fox, 1997). The loose material accumulates through the processes of weathering, erosion and deposition/sedimentation. Solidification takes place by a process called lithification/diagenesis, which involves the compaction, cementation and recrystallisation of sediments that are deeply buried (±3 km). Sedimentary rocks are also formed by the lithification of chemical precipitates that accumulate as layers of microcrystals on lake floors or subterranean cavities (Thompson & Turk, 2007; Carlson et al., 2008).
- **Metamorphic rocks:** formed by the exposure of rocks to high temperature and/or pressures during magmatic and/or tectonic events (Chernicoff & Fox, 1997). Heat from nearby magmatic intrusions and pressure induced by mountain-building and other tectonic processes causes reactions and recrystallisation of minerals resulting in new sets of minerals within metamorphosed rocks. The process of recrystallisation occurs in the solid state, or in extreme cases, in a partially molten state.

The three rock groups are characterised by important differences in the types of minerals and their textural relationships. These differences are manifest in the physical properties of the rocks, such as strength and elasticity ratios, that affect their behaviour during comminution. Consequently, the physical properties of minerals are not the only controlling factors on the effectiveness of comminution, but more importantly, the mineral assemblage and texture of the rock, which is the reason why three different rocks have been selected for trial in this research (dolerite, sandstone and marble – see Table 1 for selected characteristics of these rocks). The same is true for different types of ores where the textures ultimately determine the grain-size to which an ore needs to be milled before liberation is properly effected.

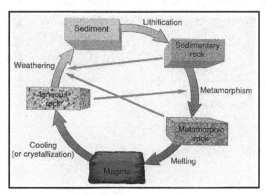

Fig. 3. The rock cycle (Chernicoff & Fox, 1997)

Dolerite is an igneous intrusive rock with medium-sized grains (Aloian, 2010), which was injected as a fluid into older water-laid sedimentary rocks (Leaman, 1973). It was injected under pressure head of at least 700 kg/cm² and caused severe disruptions of pre-existing geological features. As the initially hot fluid was forced into cold sediments, severe thermal gradients were produced during the period of injection and cooling. Dolerite is primarily composed of two essential minerals, termed plagioclase and pyroxene. The mineralogy of dolerite is closely related to the form of the intrusion and the thermal history.

Sandstones are types of sedimentary rocks, which consist particularly of mineral grains, deposited in parallel layers, which have subsequently been cemented together. Sandstones are mostly white, light grey, buff, reddish or yellowish brown in colour. Quartz is the predominant mineral found in most sandstones being chemically stable and physically durable under most weathering and transporting processes (Evans, 1972; Dietrich & Skinner, 1979). Sedimentary rocks formed by the deposition of mineral grains are classified on the basis of grain-size. Sandstones have grain-sizes ranging from 0.0625 mm to 2 mm and generally comprise quartz with or without feldspar and other mineral fragments. In addition, sandstones may also have interstitial finer grained material such as clay or cementing material which can vary in amount as a percentage of the total rock (< 5 - 25%) (McGeary et al., 2001).

Marble is an example of a metamorphic rock consisting primarily of calcite and/or dolomite. Marble may be snow white, grey, black, buff, yellowish, chocolate, pink, mahogany-red, bluish, lavender or greenish in colour. The grains within marble tend to be of a rather uniform size (Dietrich & Skinner, 1979). Marble forms by the metamorphism of limestone, during which recrystallisation of calcite occurs (Evans, 1972). Completely recrystallised limestone can result in a rock with interlocking calcite crystals and the obliteration of the stratification and other textural characteristics of the parent limestone. Impure parent limestone produces marble that contains other minerals in addition to calcite, with the most common being quartz, anorthite, serpentine, tremolite, diopside, and forsterite. The minerals present depend on the nature of the impurities and the grade of metamorphism.

Characteristic	Dolerite	Sandstone	Marble
Rock type	Igneous (hypabysal)	Sedimentary	Metamorphic
Texture	Medium grained and smooth	Medium grained and rough	Coarse grained
Principal minerals	Plagioclase and Pyroxene	Quartz	Calcite and dolomite
Principal mineral hardness	Plagioclase: 6; Pyroxene: 6	7	3
Principal mineral breakage	Two good cleavages at 90°	Fracture	Three good cleavages at 75°/105°
Specific gravity	3.00 – 3.05	2.00 – 2.60	2.6 – 2.86
Resistivity (Ω.m)	20 – 200	8 – 4 000	100 – 250 000 000
Colour	Dark bluish, weathers to brown	White, light grey, buff, reddish or yellowish brown	White, grey, black, buff, yellowish, chocolate, pink, mahogany-red, bluish, lavender or greenish
Porosity	0.1 – 0.5%	5.0 – 25.0%	0.5 – 2.0%

Table 1. Selected characteristics of dolerite, sandstone and marble

The rocks discussed above were selected for this research and were labelled with the text JS (representing James Swart) followed by an alphabetical label as shown in Table 2. This was done to prevent confusion between the different samples which were treated with electrical RF power.

Sample code	Rock family	Rock type
JSA	Igneous	Dolerite
JSB	Metamorphic	Marble
JSD	Sedimentary	Sandstone

Table 2. Rock samples chosen for this research

4. Electrical treatment techniques

Electrical treatment techniques refer to the use of electrical energy in specific ways to achieve desired changes in certain solid and liquid materials. Four specific electrical techniques currently employed include:

- Microwave pre-treatment;
- Ultrasound pre-treatment;
- High voltage electrical pulses; and
- Radio-frequency power.

4.1 Microwave pre-treatment

Numerous studies have shown that microwave pre-treatment is beneficial for:

- Drying of raisins (Kostaropoulos & Saravacos, 1995);
- Accelerating enzymatic hydrolysis of chitin (Roy et al., 2003);
- Improved grindability and gold liberation (Amankwah et al., 2005);
- Improving the moisture diffusion coefficient of wood (Li et al., 2005);
- Enhancement of phosphorus release from dairy manure (Pan et al., 2006);
- Strength reduction in ore samples (Jones et al., 2007);
- Enhancing enzymatic digestibility of switchgrass (Hu & Wen, 2008);
- A higher extractive yield of vegetable oil from Chilean hazelnuts (Uquiche et al., 2008); and
- The liberation of copper carbonatite ore after milling (Scott et al., 2008).

Microwave pre-treatment is found in many other applications where microwaves induce transient motions of free or bound charges, such as electrons or ions or charge complexes such as permanent dipoles. The resistance to these motions causes losses, which result in attenuation of the electric field and increased dissipation of energy in the material (Amankwah et al., 2005).

The most important early work on microwave pre-treatment was that of Chen et al. (1984), who investigated the reaction of 40 minerals to microwave exposure in a waveguide applicator which allowed the mineral samples to be inserted in an area of known high electric field strength. This study showed that microwave heating is dependent on the composition of the minerals.

Walkiewicz et al. (1988) later published data on microwave heating of a number of minerals and speculated on the potential reduction in grinding energy required for minerals with stress fractures induced by microwave heating. Kingman et al. (2004) published an article stating that for the first time microwave-assisted comminution may have the potential to become economically viable. This conclusion was based on significant reductions in strength, coupled with major improvements in liberation of valuable minerals.

The microwave heating system is made up of four basic components: power supply, magnetron, cavity for the heating of the target material and waveguide for transporting microwaves from the generator to the cavity. Commonly, an industrial size microwave heating system is set to a frequency of 915 MHz with a magnetron as high as 75 kW power and an average working life of 6000 hours (Smith, 1993).

Microwave heating is a sophisticated electroheat technology requiring specialist knowledge and expensive equipment if meaningful results are to be obtained (Bradshaw et al., 1998). Included in this is the precision involved in the design and construction of the magnetron and cavity.

4.2 Ultrasound pre-treatment

The use of ultrasound pre-treatment has been applied to:

- Accelerate the anaerobic digestion of sewage sludge (Tiehm et al., 1997);

- Comminution (Gaete-Garretón et al., 2000);
- Titanium tanning of leather (Peng et al., 2007);
- Ammonia steeped switchgrass for enzymatic hydrolysis (Montalbo-Lomboy et al., 2007);
- Two-Minute skin anaesthesia (Spierings et al., 2008); and
- Cassava chip slurry to enhance sugar release for subsequent ethanol production (Nitayavardhana et al., 2008).

The feasibility of the application of ultrasound energy to the grinding process as a viable avenue of study was stated at a meeting of the International Comminution Research Association in Warsaw, 1993 (Gaete-Garretón et al., 2000). One of the most significant reasons for this proposition originated in the accepted fact that inside any material there are a number of inherent cracks and ultrasonic energy has the capacity to produce crack propagation from within the particle to its outer surface, in spite of the very low energy producing an efficient fracture. An ultrasonic grinding machine can be designed in the form of a roller mill constructed over a specially designed ultrasonic transducer (Gaete-Garretón et al., 2003).

Gärtner (1953) was probably the first researcher to have attempted using ultrasonic waves in the fragmentation of particles, obtaining poor results. Leach and Rubin (1988) studied the fragmentation of resonant rocks samples fixed to the tip of an ultrasonic transducer, observing a preferred fracture at the nodes. Yerkovic et al. (1993) made grinding tests comparing standard copper ore with ultrasonic pre-treated samples in a ball mill. The pre-treated ore exhibited a 32% higher grinding rate.

An active roll, which is itself an ultrasonic transducer, is located in front of a passive roll. The vibration in extensional mode combines compression and shear action of the active roll on the mill feed. A funnel feeds the material into the gap by gravity which are then nipped by the rolls. A spring system furnishes the stress applied to the ore and the stress level can be varied by adjusting the spring tension. The rotation of the roll is produced by a variable speed electric motor. The ground ore is collected under the rolls in an iron receiver fed by gravity.

It is evident from the above description of the ultrasound mill that many different parts have to work together in the application of an ultrasonic field in the stressing zone of the material. This setup proves to be very precise and time consuming.

4.3 High voltage pulsed power

High voltage pulsed power has been applied to:

- Enhance coal comminution and beneficiation (Touryan & Benze, 1991);
- Mineral liberation (Andres et al., 2001);
- Metal peening (Zhang & Yao, 2002);
- Rock fragmentation (Cho et al., 2006);
- Recover ferrous and non-ferrous metals from slag waste (Wilson et al., 2006); and
- Convective drying of raisins (Dev et al., 2008).

The history of high voltage pulsed power can be traced back to 1752 when Benjamin Franklin discovered that lightning was a discharge of static electricity (Staszewski, 2010). It was reported that he raised a kite (with a key attached to his end of the string) which was tied to a post with a silk thread. As time passed, Franklin noticed the loose fibres on the string stretching out; he then brought his hand close to the key and a spark jumped the gap. This electrical discharge across a gap would prove significant in the research of high voltage pulsed power techniques.

In 1924 Erwin Marx described an apparatus, which produced high voltage pulses, and became known as the Marx-Generator (Fontana, 2004). It is a clever technique for generating high-voltage short-duration waveforms by charging a number of capacitors in parallel, then quickly discharging them in series. While originally based upon the use of air-dielectric spark gaps to provide the switching mechanism, solid-state variants utilizing avalanche diodes or other solid-state switching devices have been used to generate nanosecond duration pulses having amplitudes exceeding several thousand volts of direct current (Baker & Johnson, 1993).

There has been intense interest for the last several decades in the use of high-voltage pulse technology for rocks disintegration (Cho et al., 2006). The methods of electric pulse disintegration are mainly electrohydraulics and internal breakdown inside bulk solid dielectrics (Owada et al., 2003). The first method refers to the generation of an intense shock wave in water from the passage of electrical current through water and the crushing and subsequent constituent separation by the impact of that shock wave on the sample. The second method refers to the passage of electrical current through the rock and the separation of the mineral contents from the rock matrix by preferential current flow along the mineral/rock boundary interface. Rock disintegration using the second method consumes substantially less energy than that using the first method and enhanced effect of liberation of mineral constituents of rock aggregates.

A major limiting factor to spark-gap switches used in high voltage pulsed power applications was their short lifetime (Winands et al., 2005). Other shortcomings with spark gaps are related to their limited pulse repetition rate, strong electrode erosion, insulator degradation, high arc inductance, limited hold-off voltage, and costly triggering.

4.4 Radio-frequency power

The application of electrical energy in the RF heating of various materials has been successfully employed in the following:

- Electrical heating along with RF heating was used in the 1970s for the recovery of bitumen from tar sand deposits (Kawala & Atamanczuk, 1998);
- RF treatments can potentially provide an effective and rapid quarantine security protocol against codling moth larvae in walnuts as an alternative to methyl bromide fumigation (Wang et al., 2001);
- RF heating was successfully used to increase the temperature of human blood without incurring cell destruction (Pienaar, 2002);
- Treating fruit in immersion water of selected salt concentration and RF power may be used to develop an effective alternative quarantine method for fruit (Ikediala et al., 2002);
- RF power in conjunction with conventional hot water treatment can be used to develop feasible heat treatments to combat codling moths in apples (Wang et al., 2006);

- RF-based dielectric heating was used in the alkali pre-treatment of switchgrass to enhance its enzymatic digestibility (Hu et al., 2008); and
- Dielectric heating of soil using radio waves (RW) can be applied to support various remediation techniques, namely biodegradation and soil vapor extraction, under in situ or ex situ conditions (Roland et al., 2008).

Dielectrics have two important properties (Jones et al., 2002):

- They have very few free charge carriers. There is very little charge carried through the material matrix when an external electrical field is applied.
- The molecules or atoms comprising the dielectric exhibit a dipole movement.

The principle of dielectric heating basically involves the absorption of energy by dipoles (Chee et al., 2005). A dipole is essentially two equal and opposite charges separated by a finite distance. An example of this is the stereochemistry of covalent bonds in a water molecule, giving the water molecule a dipole movement. Water is the typical case of a non-symmetric molecule. Dipoles may be a natural feature of the dielectric or they may be induced (Kelly & Rowson, 1995). Distortion of the electron cloud around non-polar molecules or atoms through the presence of an external electric field can induce a temporary dipole movement. This movement generates friction inside the dielectric and the power is dissipated subsequently as heat. The interaction of dielectric materials with electromagnetic radiation in a given frequency band results in energy absorbance (Wang et al., 2001; Jones et al., 2002). The power coupled into a sample is nearly constant when the electric field intensity and dielectric loss factor do not vary at a given frequency. The heat generated per unit volume (P in W/m^3) in a dielectric material when exposed to RF power can be expressed as (Nelson, 1996):

$$P = 5.56 \times 10^{-11} \times f \times E^2 \times \varepsilon \qquad W/m^3 \qquad (1)$$

Where

$f \equiv$ frequency of radiation in Hertz (Hz)
$E \equiv$ the electric field intensity in Voltage per meter (V/m)
$\varepsilon \equiv$ the permittivity of the material

Moreover, the amount of heat (Q) required to change the temperature of a given material is proportional to the mass of the material and to the temperature change as given by Giancoli (2005):

$$Q = C \times m \times \Delta T \qquad J \qquad (2)$$

Where

$\Delta T \equiv$ temperature change in degrees Celsius (°C)
$m \equiv$ the sample mass in kilogram (kg)
$C \equiv$ specific heat capacity in Joules per kilogram per degrees Celsius (J/kg/°C)

Subsequently, temperature rise within the sample due to absorbed electromagnetic energy is really a function of the heating time. The temperature increase can be estimated by assuming that the electric field is uniform and the dielectric properties are relatively constant. The temperature increase (ΔT in °C) of the sample during RF heating can furthermore be expressed as (Halverson et al., 1996):

$$\Delta T = \frac{k \times P}{C \times m} \times \Delta t \quad °C \tag{3}$$

Where

k ≡ coupling coefficient
P ≡ input power (W)
Δt ≡ RF heating time in seconds (s)

The practical setup used to achieve the transfer of RF power to a dielectric sample is shown in Fig. 4, and consists of a transformer, rectifier, oscillator, an inductance-capacitance pair commonly referred to as the 'tank circuit', and the work circuit (Wang et al., 2001). The transformer raises the voltage to 9 kV and the rectifier provides a direct current which is then converted by the oscillator into RF power at 27 MHz. This frequency is determined by the values of the inductance and capacitor in the tank circuit. The parallel-plate electrodes, with sample in-between, act as the capacitor in the work circuit. The gap of the electrode plates can be changed to adjust RF power coupled to the sample between the two plates.

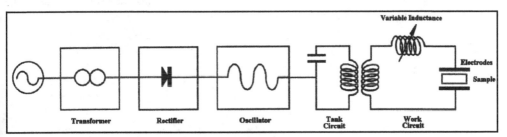

Fig. 4. Practical setup used in the dielectric heating of a material (Wang et al., 2001)

Three (microwave, ultrasound and high voltage pulsed power) of the four electrical treatment techniques noted above have been successful in weakening the mineral grain boundaries of rocks, thereby enhancing mineral liberation within the rock comminution process. This is accomplished by the generation of stress within the material, which gives rise to fractures and breakages. The weakening of mineral grain boundaries may yet be achieved by using RF power.

5. Proposed new treatment: RF power

Emanating from the above scientific literature on the use of electrical energy in various treatment techniques, the following hypotheses are made:

- The successful transfer of RF power to specific rocks through dielectric heating may exhibit positive effects on the textural characteristics of these samples; and
- These textural changes may further contribute to enhancing the rock comminution process, thereby increasing the percentage of valuable liberated minerals.

As far as could be established, no current literature exists substantiating these hypotheses. A novel electrical treatment technique of rock samples involving RF power within the very-high frequency (VHF) range (30-300 MHz) is subsequently presented.

A high power RF amplifier may be connected to a rock sample (acting as a dielectric material) by means of a suitable coupling device. RF power is transferred from the amplifier to the rock sample at the resonating frequency. Confirmation of power transfer may be determined through the following results:

- Temperature increase on the surface of the rock sample;
- Surface colour change of the sample;
- Screening of particles from pre-treated and non-treated sample;
- Scanning electron microscope (SEM) analysis of pre-treated and non-treated samples; and
- Power consumption analysis of pre-treated and non-treated samples in a ball mill.

The practical setup of this experiment is shown in Fig. 5. A commercial RF transceiver (ICOM IC-V8000) may be used in conjunction with two RF amplifiers (MIRAGE PAC30-130B) to generate the power required at the resonating frequency. However, the output impedance of the RF amplifiers is 50 Ω while the input impedance of the rock samples may vary dramatically from a few hundred ohm to a few thousand ohm (determined by a practical setup discussed by Swart et al. (2005)).This necessitates the use of a matching network and a coupling device, which are presented next.

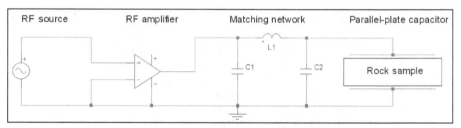

Fig. 5. Practical setup of the experiment based on research by Wang et al. (2001)

6. Innovative coupling technique

Two RF measurement coupling techniques currently exist for connecting RF equipment to dielectric materials, namely the cylindrical capacitor with coaxial electrodes and the parallel-plate capacitor (PPC) with disk electrodes.

Numerous articles list the usefulness of the **cylindrical capacitor** in measuring electrical impedances (Levitskaya & Sternberg, 2000; Bagdassarov & Slutskii, 2003; Azimi & Golnabi, 2009). A cylindrical capacitor consists of a three-part coaxial capacitance sensor in which the middle one acts as the main sensing probe (Azimi & Golnabi, 2009). The outer conductor is considered to be a guard ring in order to reduce stray capacitance and error measurements. Aluminium material is often used for manufacturing the capacitor tube electrodes (Rutschlin et al., 2006). The cylindrical capacitor extends the frequency limit of measurements to 1 GHz for materials with a dielectric permittivity of less than 25 (Levitskaya & Sternberg, 2000). However, cutting a sample of marble (dielectric permittivity of 8) into a cylindrical form with exact diameter spacing proves cumbersome and difficult in the absence of a core-drill. For this reason the PPC was reviewed as a coupling device.

A **PPC** with disk electrodes is formed when a dielectric material or sample is sandwiched between two conducting plates (see Fig. 6). These conducting plates (made from copper due to its good conductivity (Zaghloul, 2008) are connected to relevant test equipment via standard RF connectors and coaxial cables. Swart et al. (2009) suggests that high frequency measurements (up to 950 MHz) using PPCs are easier on dielectric material samples with small widths (<5 mm) and low dielectric permittivities (around 8). Subsequently, all rock samples were cut to a width of 4 mm (using a laboratory rock cutter) and sandwiched within a PPC connected to a matching network housed in a novel wooden jig.

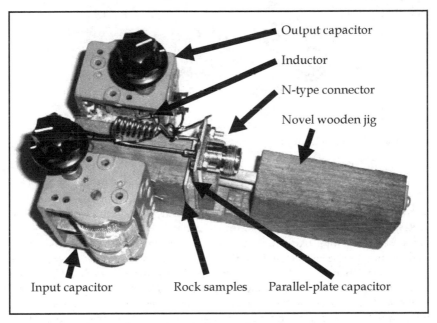

Output capacitor

Inductor

N-type connector

Novel wooden jig

Input capacitor   Rock samples   Parallel-plate capacitor

Fig. 6. Two rock samples inside a PPC with the matching network

The matching network ensures maximum power transfer to the rock samples and was made from mechanical plate trimmer capacitors (5-100 pf) and an inductor constructed from silver wire (silver solder). Silver has a lower resistivity (1.624×10^{-8} Ω-cm) than that of copper (1.728×10^{-8} Ω-cm) at 20 °C and is therefore a better conductor with less attenuation (Hutchinson, 2001). Hence, it will not heat up as quickly as copper will, which could weaken the soldering joints. Pozar (2005) substantiates this claim by noting that the conductivity of silver (6.173×10^{7} S/m at 20 °C) is higher than that of copper (5.813×10^{7} S/m at 20 °C).

7. Practical setup of the equipment

The matching network's performance was evaluated using two RF amplifiers (MIRAGE PAC30-130B) driven by a commercial RF transceiver (ICOM IC-V8000). The RF transceiver generated a 3.2 W RF signal which was amplified by the first RF amplifier to approximately

32 W, and in turn, to approximately 113 W by the second RF amplifier. This was necessary because the RF transceiver was not capable of providing more than 70 W of RF power. The input to the RF amplifiers was limited to 35 W to ensure correct operation of the driver stages. The practical setup is shown in Fig. 7.

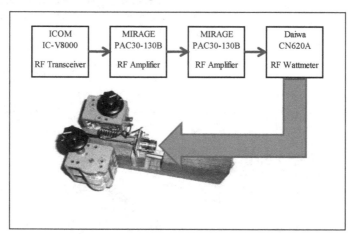

Fig. 7. Practical setup to determine the efficacy of the matching network

The output of the RF transceiver was first connected straight to the matching network through a RF wattmeter to determine the standing wave ratio (SWR) of the circuit. The reason for this is to ensure that the SWR value remain as close as possible to one, in order to prevent an excess of reflected power damaging the output stage of the RF amplifier. With the two RF amplifiers bypassed, the RF transceiver was activated (keyed) to generate a 3.2 W signal at 160 MHz (approximate centre of the VHF range which extends from 30 – 300 MHz). The trimmer capacitors were then fine-tuned to obtain the lowest possible SWR. The RF amplifiers were then switched on (thus connecting the amplifiers directly into the circuit between the RF transceiver and the matching network). The RF transceiver was keyed again and approximately 113 W of forward power was measured with the wattmeter. Two different wattmeters were used to verify the reliability of the measurements. The reflected power measured approximately 1.8 W resulting in a SWR reading of 1.308. The successful transfer of RF power (82 W) to the rock samples was further collaborated by a significant rise in surface temperature, as described in the following results section.

8. Results of treating rock sample with RF power

Rock samples of specific size which have been exposed to a known amount of RF power at a given frequency are referred to as treated samples, while those which were not exposed are termed untreated samples. Possible changes that were considered include textural, phase, grindability, colour, temperature and electrical property changes. Textural changes (changes in grain size and inter-grain boundary relationships) were considered using polarizing optical microscopy on polished thin sections of the rock samples. Geological type samples are often cut into thin slices, using diamond type saws, and then fixed onto glass slides using a strong adhesive (Tiedt & Pretorius, 2002). The mount is then polished, and subsequently coated with a conductive layer. These thin sections were also used to determine phase changes (changes in mineral assemblage) using polarizing optical microscopy. The textural and phase changes were analyzed with a Phillips model XL30 DX41 scanning electron microscope. Grindability, being the changes in the power consumption during grinding and changes in the particle size distribution after grinding, was determined by measuring the power consumption during milling (using a HIOKI 3286-20 clamp on power meter) and by performing particle size analyses (sieve tests of screen sizes 250 µm, 150 µm, 90 µm and 38 µm). Surface colour changes were visually observed while surface temperature changes were measured (using a LUTRON TM-2000 digital thermometer and K-type thermocouples pressed firmly against the surface of the rock samples). Contrasts between the electrical properties (resonating frequencies and impedances obtained from a HP 8752C network analyzer) of the untreated and treated samples are further indicated. The results from the above considerations were interpreted in terms of the mineralogical and chemical composition of the samples.

8.1 SEM analysis of photomicrographs

A comparison of the photomicrographs of the untreated and treated rock samples reveals no significant differences in grain size, grain shape, minerals present or inter-granular textures. No visible cracks or fractures exist along the mineral grain boundaries of the treated rock samples, as shown in Fig. 8 – 10.

Fig. 8. Photomicrographs of the untreated (left) and treated (right) JSA rock sample taken under cross-polarized light

Fig. 9. Photomicrographs of the untreated (left) and treated (right) JSB rock sample taken under cross-polarized light

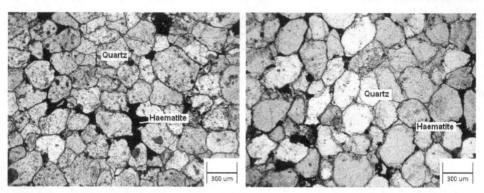

Fig. 10. Photomicrographs of the untreated (left) and treated (right) JSD rock sample taken under plane-polarized light

8.2 Particle screen analysis

Determination of the relative grindability of the untreated and treated samples was done by measuring the power consumption during grinding and comparing the particle size distribution after the grinding process. The untreated and treated samples were ground down to powder form in a laboratory swing mill, which consists of a shallow cylinder; two internal rings and a heavy disc. These mills are designed for reduction of materials to extremely fine powders for preparation of samples for spectra analysis. All samples were milled for 2 minutes with corresponding power measurements taken of the power consumed. A small brush was used to clean out the grounded samples (in the form of powder or dust) from the pot, which were then weighed with a digital scale.

The powder samples were next transferred to particle screening sieves (250 μm, 150 μm, 90 μm and 38 μm screens placed on top of each other). This screen combination was placed in an ENDECOTTS EFL2000 shaker for 5 minutes. Rock sample particles left behind in each screen was weighed individually. These weightings were converted into percentages by dividing each weighting by the total mass and cumulative mass percentages by adding successive mass percentages. The results of this evaluation are shown in Fig. 11 – 13, where the untreated samples are shown by means of a triangle or cross. The treated samples are indicated by means of a diamond or square. The left sketch indicates the particle size distribution to cumulative mass, while the right hand sketch shows the frequency of occurrence for each grain size. Four samples were used in the analysis of the JSA and JSD rock samples, to ensure repeatability and reliability of the results. Three JSB rock samples were used in this analysis, as it showed little or no variation in post-grinding particle size distribution between the untreated and treated rock samples.

However, JSA and JSD revealed minor to major variations. The treated JSA sample (right hand graph) shows a significant coarser grain size distribution with a mode value of 90 μm, whereas it is 38 μm for the untreated samples. Similarly, the d_{80} (nominal sieve size allowing 80% of the powered sample to pass through – left hand gragh) is approximately 85 μm for the treated ones, but less than 38 μm for the untreated samples. This means that for the same amount of grinding (2 minutes) the treated samples were reduced in size to a lesser extent than the untreated samples, suggesting reduced grindability. This may also indicate that fewer fines (smaller particles) are generated and therefore over-grinding is reduced. A similar situation is evident for the JSD sample, which is a sandstone with granular textures in which sand grains are cemented with matrix material such as haematite, whereas the JSA sample has a typical igneous texture of interlocking crystals.

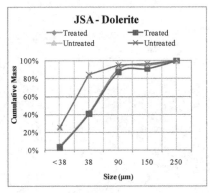 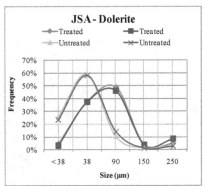

Fig. 11. Particle screen results for the untreated and treated JSA sample

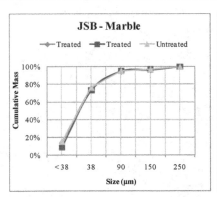 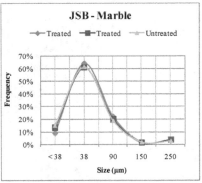

Fig. 12. Particle screen results for the untreated and treated JSB sample

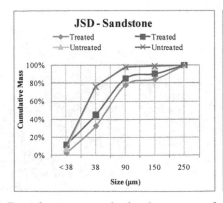 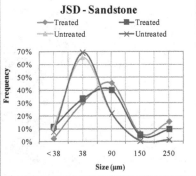

Fig. 13. Particle screen results for the untreated and treated JSD sample

Polished sections of the powdered samples were obtained to check for textural changes with regard to particle sizes of the untreated and treated samples (see Fig. 14 - 16). The size of the mineral grains is indicated, revealing no significant reduction in size between the untreated and treated samples.

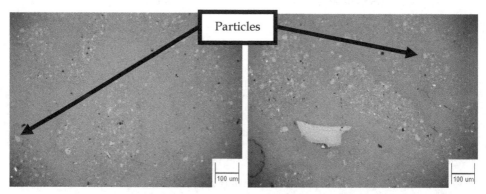

Fig. 14. Photomicrographs of the untreated (left) and treated (right) JSA sample

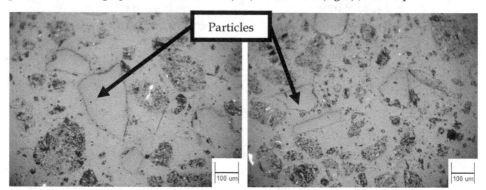

Fig. 15. Photomicrographs of the untreated (left) and treated (right) JSB sample

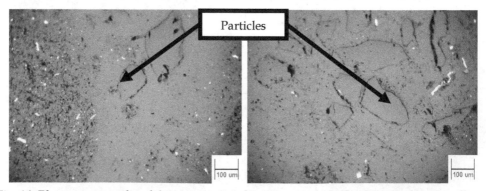

Fig. 16. Photomicrographs of the untreated (left) and treated (right) JSD sample

8.3 Electrical properties and surface colour changes

The transfer of RF power to the rock samples resulted in a surface temperature rise due to RF heating of the dielectric material. Table 3 presents visual effects of RF heating on the three rock samples (JSA, JSB and JSD), as well as the temperature reached after a specified time of RF treatment.

Sample code and type	Maximum temperature (degrees Celsius)	Time (min)	Screen change	Colour change	Untreated sample	Treated sample
JSA 30 x 19 x 4 mm Dolerite Igneous	151	4	Yes	No		
Frequency / resistance					160.14 MHz / 1264 Ω	164.37 MHz / 1319 Ω
Power consumption					1419 W	1420 W
JSB 31 x 19 x 4 mm Marble Metamorphic	107	2	No	Yes		
Frequency / resistance					162.58 MHz / 1325 Ω	161.57 MHz / 1467 Ω
Power consumption					1357 W	1356 W
JSD 30 x 19 x 4 mm Sandstone Sedimentary	55	5	Yes	No		
Frequency / resistance					170.08 MHz / 1169 Ω	169.86 MHz / 1292 Ω
Power consumption					1357 W	1358 W

Table 3. Surface colour changes and maximum temperatures reached for samples JSA, JSB and JSD (untreated on the left and treated on the right)

Colour changes and maximum temperature reached with 82 W of RF power at 160 MHz is indicated. S-parameters (in the form of Cartesian Coordinates) for the rock samples were obtained from a network analyser, from which the resonating frequencies and resistances were calculated for the untreated and treated rock samples. No significant variations were observed between the untreated and treated samples with regard to two of their electrical properties, being the resonating frequencies and resistances. The power consumption (sum of the power consumed by the RF equipment and milling machine) of the untreated and treated samples also reveals no significant variations.

9. Conclusion

One of the primary aims of this research was to design and develop a suitable coupling device to connect relevant electronic equipment (test instruments and amplifiers) to various rock samples. Maximum power transfer to the rock sample at a specific frequency was achieved with the use of a PPC and matching network housed in a novel wooden jig. Inserting specific sized rock samples into this coupling device proved simple and effective, being neither time consuming or difficult.

Only two of the three samples (JSA and JSD) revealed a notable change in their particle size distribution. The fact that the percentage of larger sized particles increased (from 38 μm to 90 μm) suggests that the rock was **strengthened** rather than weakened (see Fig. 11 and Fig. 13). A possible application could be the prevention of over-grinding during comminution, which may have benefits during mineral processing. Moreover, RF power could further be used in the colouring of rock surfaces (see Table 3).

Evaluating the effects of RF power treatment on rocks has brought to light that mineral grain boundaries within specified rock samples are not significantly weakened by RF treatment. This was firstly confirmed by the similar electrical properties of the untreated and treated samples, where consistent values for the resonating frequency were obtained from a network analyser. Secondly, the SEM analysis of the untreated and treated rock samples revealed no significant changes in the form of fractures or breakages along the mineral grain boundaries. Photomicrographs of the thin sections of the two rock samples were used in this analysis. The particle size distribution of both samples further revealed no weakening or softening of the rock, as the percentage of smaller sized particles did not increase in the treated samples. It may therefore be stated that treating rock samples with RF power within the VHF range will not significantly improve rock comminution and mineral liberation.

10. Acknowledgment

I acknowledge Prof. Christo Pienaar (Director of the Telkom Centre of Excellence) for his specific guidance relating to the methodology employed in this research. I also would like to acknowledge Prof. Peter Mendonidis (Principal Lecturer in Metallurgical Engineering at Vaal University of Technology) for explaining difficult terms and principles relating to rock comminution. I further acknowledge my colleague, Ruaan Schoeman, who often provided a listening ear to my concerns and battles regarding this research. Appreciation is also extended to the Central Research Committee at VUT for their financial contribution to this research.

11. References

Aloian, M. (2010). *What are igneous rocks*. New York: Crabtree Publishing Company.

Amankwah, R. K., Khan, A. U., Pickles, C. A., & Yen, W. T. (2005). Improved grindability and gold liberation by microwave pretreatment of a free-milling gold ore. *International Journal of Mineral Processing and Extractive Metallurgy Review*, Vol. 114, No. C, pp. 30-36, ISSN 0882-7508

Andres, U., Timoshkin, I., & Soloviev, M. (2001). Energy consumption and liberation of minerals in explosive electrical breakdown of ores. *Mineral Processing and Extractive Metallurgy: Transactions of the Institute of Mining and Metallurgy, Section C*, Vol. 110, No., pp. 149-157, ISSN 0371-9553

Azimi, P., & Golnabi, H. (2009). Precise formulation of electrical capacitance for a cylindrical capacitive sensor. *Journal of Applied Science*, Vol., No. 9, pp. 1556-1561, ISSN 1812-5654

Bagdassarov, N. S., & Slutskii, A. B. (2003). Phase transformations in calcite from electrical impedance measurements. *Phase Transitions: A Multinational Journal*, Vol. 76, No. 12, pp. 1015 - 1028, ISSN 0141-1594

Baker, R. J., & Johnson, B. P. (1993). Applying the Marx bank circuit configuration to power MOSFETs. *Electronics Letters*, Vol. 29, No. 1, pp. 56-57, ISSN 0013-5194

Best, M. G., & Christiansen, E. H. (2001). *Igneous Petrology*. Massachusetts: Blackwell Science.

Bradshaw, S. M., van Wyk, E. J., & de Swardt, J. B. (1998). Microwave heating principles and the application to the regeneration of granular activated carbon. *The Journal of The South African Institute of Mining and Metallurgy*, Vol. July/August, No., pp. 201-210, ISSN 0038-223X

Carlson, D. H., Plummer, C. C., & McGeary, D. (2008). *Physical Geology: Earth Revealed* (7th ed.). New York: McGraw-Hill.

Chee, S. N., Johansen, A. L., Gu, L., Karlsen, J., & Heng, P. W. S. (2005). Microwave Drying of Granules Containing a Moisture-Sensitive Drug: A Promising Alternative to Fluid Bed and Hot Air Oven Drying. *Chemical & Pharmaceutical Bulletin*, Vol. 53, No. 7, pp. 770-775, ISSN 0009-2363

Chen, T. T., Dutrizac, J. E., Haque, K. E., Wyslouzil, W., & Kashyap, S. (1984). The relative transparency of minerals to microwave radiation. *Canadian Metallurgical Quarterly*, Vol. 23, No. 3, pp. 349-351, ISSN 0008-4433

Chernicoff, S., & Fox, H. A. (1997). *Essential of Geology*. New York: Worth Publishers.

Cho, S. H., Mohanty, B., Ito, M., Nakamiya, Y., Owada, S., Kubota, S., et al. (2006). *Dynamic fragmentation of rock by high-voltage pulses*. Paper presented at the The 41st U.S. Symposium on Rock Mechanics (USRMS), Golden, Colorado. June 17-21

De Waal, P. (2007). *Tomorrow's Technology - Out of Africa - Today*. Paper presented at the The Fourth Southern African Conference on Base Metals, Swakopmund, Namibia. 23-25 July

Dev, S. R. S., Padmini, T., Adedeji, A., Gariepy, Y., & Raghavan, G. S. V. (2008). A comparative study on the effect of chemical, microwave, and pulsed electric pretreatments on convective drying and quality of raisins. *Drying Technology*, Vol. 26, No. 10, pp. 1238-1243, ISSN 07373937

Dietrich, R. V., & Skinner, B. J. (1979). *Rocks and rock minerals*. New York: John Wiley and Sons.

Evans, I. O. (1972). *Rocks, Minerals & Gemstones*. London: The Hamlyn Publishing Group.

Fontana, R. J. (2004). Recent System Applications of Short-Pulse Ultra-Wideband (UWB) Technology. *IEEE Transactions on Microwave Theory and Techniques,* Vol. 52, No. 9, pp. 2087-2104, ISSN 0018-9480

Gaete-Garretón, L., Vargas-Hernandez, Y., Chamayou, A., Dodds, J. A., Valderama-Reyes, W., & Montoya-Vitini, F. (2003). Development of an ultrasonic high-pressure roller press. *Chemical Engineering Science,* Vol. 58, No. 19, pp. 4317-4322, ISSN 0009-2509

Gaete-Garretón, L. F., Vargas-Hermández, Y. P., & Velasquez-Lambert, C. (2000). Application of ultrasound in comminution. *Ultrasonics,* Vol. 38, No. 1-8, pp. 345-352, ISSN 0041-624X

Gärtner, W. (1953). Über die Möglichkeit der zerkleinerung suspendierter stoffe durch ultrashall. *Acustica,* Vol. 3, No., pp. 124-128, ISSN 1610-1928

Giancoli, D. C. (2005). *Physics - principles with applications* (6th Ed ed.). New Jersey: Pearson Prentice Hall.

Halverson, S. L., Burkholder, W. E., Bigelow, T. S., Norsheim, E. V., & Misenheimer, M. E. (1996). High-power microwave radiation as an alternative insect control method for stored products. *J. Econ. Entomol,* Vol. 89, No., pp. 1638-1648, ISSN 0022-0493

Henan Chuangxin Building-material Equipment Co. (2009). Retrieved 6 November 2009, from http://www.enchuangxin.com/Mineral%20Processing%20Equipment/

Hu, Z., Wang, Y., & Wen, Z. (2008). Alkali (NaOH) pretreatment of switchgrass by radio frequency-based dielectric heating. *Applied Biochemistry and Biotechnology,* Vol. 148, No. 1-3, pp. 71-81, ISSN 0273-2289

Hu, Z., & Wen, Z. (2008). Enhancing enzymatic digestibility of switchgrass by microwave-assisted alkali pretreatment. *Biochemical Engineering Journal,* Vol. 38, No. 3, pp. 369-378, ISSN 1369-703X

Hutchinson, C. (Ed.). (2001). *The ARRL Handbook for Radio Amateurs* (78th ed.). Newington: ARRL.

Ikediala, J. N., Hansen, J. D., Tang, J., Drake, S. R., & Wang, S. (2002). Development of a saline water immersion technique with RF energy as a postharvest treatment against codling moth in cherries. *Postharvest Biology and Technology,* Vol. 24, No., pp. 209-221, ISSN 0925-5214

Jones, D. A., Kingman, S. W., Whittles, D. N., & Lowndes, I. S. (2007). The influence of microwave energy delivery method on strength reduction in ore samples. *Chemical Engineering and Processing,* Vol. 46, No. 4, pp. 291-299, ISSN 0255-2701

Jones, D. A., Lelyveld, T. P., Mavrofidis, S. D., Kingman, S. W., & Miles, N. J. (2002). Microwave heating applications in environmental engineering--a review. *Resources, Conservation and Recycling,* Vol. 34, No. 2, pp. 75-90, ISSN 0921-3449

Kawala, Z., & Atamanczuk, T. (1998). Microwave-enhanced thermal decontamination of soil. *Environmental Science and Technology,* Vol. 32, No. 17, pp. 2602-2607, ISSN 0013-936X

Kelly, R. M., & Rowson, N. A. (1995). Microwave reduction of oxidised ilmenite concentrates. *Minerals Engineering,* Vol. 8, No. 11, pp. 1427-1438, ISSN 0892-6875

King, R. P. (2001). *Modelling and simulation of mineral processing systems.* Oxford: Butterworth-Heinemann.

Kingman, S. W., Jackson, K., Cumbane, A., Bradshaw, S. M., Rowson, N. A., & Greenwood, R. (2004). Recent developments in microwave-assisted comminution. *Int. J. Min. Proc,* Vol. 74, No. 1-4, pp. 71-83, ISSN 1478-6478

Klein, C. (2002). *The 22nd edition of the manual of Mineral Science.* New York: John Wiley & Sons.

Kostaropoulos, A. E., & Saravacos, G. D. (1995). Microwave Pre-treatment for Sun-Dried Raisins. *Journal of Food Science,* Vol. 60, No. 2, pp. 344-347, ISSN 1750-3841

Leach, M. F., & Rubin, G. A. (1988). *Fragmentation of Rocks Under Ultrasonic Loading.* Paper presented at the Ultrasonic Symp of the IEEE, Chicago, USA. 2-5 October

Leaman, D. E. (1973). *The engineering properties of Tasmanian dolerite, with particular reference to the route of the Bell Bay Railway,* Tasm. Dep. Mines. Tech. Rept.

Levitskaya, T. M., & Sternberg, B. K. (2000). Laboratory measurement of material electrical properties: extending the application of lumped-circuit equivalent models to 1 GHz. *Radio Science,* Vol. 35, No. 2, pp. 371-383, ISSN 0048-6604

Li, X., Zhang, B., Li, W., & Li, Y. (2005). Research on the effect of microwave pretreatment on moisture diffusion coefficient of wood. *Wood Science and Technology,* Vol. 39, No. 7, pp. 521-528, ISSN 0043-7719

McGeary, D., Plummer, C. C., & Carlson, D. (2001). *Physical Geology EARTH REVEALED* (4th ed.). New York: WCB/McGraw-Hill.

Montalbo-Lomboy, M., Srinivasan, G., Raman, D. R., Anex Jr, R. P., & Grewell, D. (2007). *Influence of ultrasonics in ammonia steeped switchgrass for enzymatic hydrolysis.* Paper presented at the 2007 ASABE Annual International Meeting, Technical Papers, Minneapolis, MN, United states. 17-20 June

Moran, C. (2009). *Submission to the Australian Government Energy White Paper.* Retrieved 7 July 2011. From http://www.ret.gov.au/energy/Documents/ewp/pdf/EWP%200047%20DP%20Submission%20-%20Sustainable%20Minerals%20Institute.pdf.

Nelson, S. O. (1996). Review and assessment of RF and microwave energy for stored-grain insect control. *Trans. ASAE,* Vol. 39, No., pp. 1475-1484, ISSN 0001-2351

Nitayavardhana, S., Rakshit, S. K., Grewell, D., Van Leeuwen, J., & Khanal, S. K. (2008). Ultrasound pretreatment of cassava chip slurry to enhance sugar release for subsequent ethanol production. *Biotechnology and Bioengineering,* Vol. 101, No. 3, pp. 487-496, ISSN 0006-3592

Owada, S., Ito, M., Ota, T., Nishimura, T., Ando, T., Yamashita, T., et al. (2003). *Application of electrical disintegration to coal.* Paper presented at the 22th International Mineral Processing Congress, Cape Town, South Africa. 28 September - 3 October

Pan, S., Lo, K. V., Ping, H. L., & Schreier, H. (2006). Microwave pretreatment for enhancement of phosphorus release from dairy manure. *Journal of Environmental Science and Health - Part B Pesticides, Food Contaminants, and Agricultural Wastes,* Vol. 41, No. 4, pp. 451-458, ISSN 0360-1234

Peng, B., Shi, B., Sun, D., Chen, Y., & Shelly, D. C. (2007). Ultrasonic effects on titanium tanning of leather. *Ultrasonics Sonochemistry,* Vol. 14, No. 3, pp. 305-313, ISSN 1350-4177

Perkins, D. (1998). *Mineralogy.* New Jersey: Prentice Hall.

Pienaar, H. C. (2002). *Design and development of a class E dielectric blood heater.* DTech, Vaal Triangle Technikon, Vanderbijlpark.

Pozar, D. M. (2005). *Microwave Engineering* (3rd ed.). Massachusetts: John Wiley & Sons.

Rapp, G. (2009). Properties of Minerals. In B. Herrmann & G. A. Wagner (Eds.), *Archaeomineralogy* (pp. 17-43): Springer Berlin Heidelberg.

Roland, U., Buchenhorst, D., Holzer, F., & Kopinke, F. D. (2008). Engineering Aspects of Radio-Wave Heating for Soil Remediation and Compatibility with Biodegradation. *Environmental Science & Technology,* Vol. 42, No. 4, pp. 1232-1237, ISSN 0013-936X

Roy, I., Mondal, K., & Gupta, M. N. (2003). Accelerating Enzymatic Hydrolysis of Chitin by Microwave Pretreatment. *Biotechnology Progress,* Vol. 19, No. 6, pp. 1648-1653, ISSN 8756-7938

Rutschlin, M., Cloete, J. H., & Palmer, K. D. (2006). A guarded cylindrical capacitor for the non-destructive measurement of hard rock core samples. *Measurement Science and Technology,* Vol. 17, No. 6, pp. 1390-1398, ISSN 0957-0233

Sadrai, S., Meech, J. A., Ghomshei, M., Sassani, F., & Tromans, D. (2006). Influence of impact velocity on fragmentation and the energy efficiency of comminution. *International Journal of Impact Engineering,* Vol. 33, No. 1-12, pp. 723-734, ISSN 0734-743X

Scott, G., Bradshaw, S. M., & Eksteen, J. J. (2008). The effect of microwave pretreatment on the liberation of a copper carbonatite ore after milling. *International Journal of Mineral Processing,* Vol. 85, No., pp. 121-128, ISSN 0301-7516

Shafiee, S., & Topal, E. (2009). When will fossil fuel reserves be diminished? *Energy Policy,* Vol. 37, No. 1, pp. 181-189, ISSN 0301-4215

Skinner, B. J., & Porter, S. C. (1992). *The Dynamic Earth an introduction to physical geology* (2nd ed.). New York: John Wiley & Sons.

Smith, R. D. (1993). *Large industrial microwave power supplies.* Paper presented at the Proc. Microwave-Induced reactions workshops, Pacific Grove, California. April

Spierings, E. L. H., Brevard, J. A., & Katz, N. P. (2008). Two-Minute Skin Anesthesia Through Ultrasound Pretreatment and Iontophoretic Delivery of a Topical Anesthetic: A Feasibility Study. *Pain Medicine,* Vol. 9, No. 1, pp. 55-59, ISSN 1526-2375

Staszewski, L. (2010). *Lightning Phenomenon – Introduction and Basic Information to Understand the Power of Nature.* Paper presented at the International Conference Environment and Electrical Engineering 2010, Prague, Czech Republic. 16-19 May

Swart, A. J., Pienaar, H. C. v., & Mendonidis, P. (2005). *Radio frequencies effect on rock comminution.* Paper presented at the Annual Faculty Research Seminar, Vaal University of Technology, Emfuleni Conference Centre, Vanderbijlpark. July 2005

Swart, J., Mendonidis, P., & Pienaar, C. (2009). The Electrical Properties of Chlorite Tremolite Marble measured for a range of Radio-Frequencies. *Mineral Processing and Extractive Metallurgy Review: An International Journal,* Vol. 30, No. 4, pp. 307 - 326, ISSN 0882-7508

Tarbuck, E. J., & Lutgens, F. K. (1999). *Earth An introduction to physical geology* (6th ed.). New Jersey: Prentice Hall.

Thompson, G. R., & Turk, J. (2007). *Earth Science and the Environment* (4th ed.). Belmont: Thompson Brooks/Cole.

Tiedt, L. R., & Pretorius, W. E. (2002). An introduction to electron microscopy and x-ray microanalysis [Electronic Version]. *Laboratory for Electron Microscopy*, North-West University.

Tiehm, A., Nickel, K., & Neis, U. (1997). The use of ultrasound to accelerate the anaerobic digestion of sewage sludge. *Water Science and Technology*, Vol. 36, No., pp. 121-128, ISSN 0273-1223

Touryan, K. J., & Benze, J. W. (1991). *Enhanced Coal Comminution And Beneficiation using Pulsed Power Generated Shocks*. Paper presented at the Pulsed Power Conference, 8th IEEE International, San Diego, California. 16-19 June

Tromans, D. (2008). Mineral comminution: Energy efficiency considerations. *Minerals Engineering*, Vol. 21, No. 8, pp. 613-620, ISSN 0892-6875

Uquiche, E., Jerez, M., & Ortiz, J. (2008). Effect of pretreatment with microwaves on mechanical extraction yield and quality of vegetable oil from Chilean hazelnuts (Gevuina avellana Mol). *Innovative Food Science and Emerging Technologies*, Vol. 9, No. 4, pp. 495-500, ISSN 1466-8564

Walkiewicz, J. W., Kazonich, G., & McGill, S. L. (1988). Microwave heating characteristics of selected minerals and compounds. *Minerals and Metallurgical Processing*, Vol., No., pp. 39-42, ISSN 0747-9182

Walther, J. V. (2005). *Essentials of geochemistry*. Massachusetts: Jones & Bartlett Publishers.

Wang, S., Birla, S. L., Tang, J., & Hansen, J. D. (2006). Postharvest treatment to control codling moth in fresh apples using water assisted radio frequency heating. *Postharvest Biology and Technology*, Vol. 40, No. 1, pp. 89-96, ISSN 0925-5214

Wang, S., Ikediala, J. N., Tang, J., Hansen, J. D., Mitcham, E., Mao, R., et al. (2001). Radio frequency treatments to control codling moth in in-shell walnuts. *Postharvest Biology and Technology*, Vol. 22, No., pp. 29-38, ISSN 0925-5214

Wang, Y., & Forssberg, E. (2007). Enhancement of energy efficiency for mechanical production of fine and ultra-fine particles in comminution. *China Particuology*, Vol. 5, No. 3, pp. 193-201, ISSN 1672-2515

Wenk, H. R., & Bulakh, A. (2004). *Minerals their Constitution and Origin*. Cambridge: Cambridge University Press.

Wills, B. A. (1992). *Mineral processing technology* (5th ed.). Oxford: Pergamon Press.

Wilson, M. P., Balmer, L., Given, M. J., MacGregor, S. J., Mackersie, J. W., & Timoshkin, I. V. (2006). Application of electric spark generated high power ultrasound to recover ferrous and non-ferrous metals from slag waste. *Minerals Engineering*, Vol. 19, No. 5, pp. 491-499, ISSN 0892-6875

Winands, G. J. J., Liu, Z., Pemen, A. J. M., van Heesch, E. J. M., & Yan, K. (2005). Long lifetime, triggered, spark-gap switch for repetitive pulsed power applications. *Review of Scientific Instruments*, Vol. 76, No. 8, pp. 085107-085106, ISSN 0034-6748

Wolde-Rufael, Y. (2010). Coal consumption and economic growth revisited. *Applied Energy*, Vol. 87, No. 1, pp. 160-167, ISSN 0306-2619

Woollacott, L. C., & Eric, R. H. (1994). *Mineral and metal extraction: an overview*. Johannesburg: The South African Institute of Mining and Metallurgy.

Yerkovic, C., Menacho, J., & Gaete, L. (1993). Exploring the ultrasonic comminution of copper ores. *Minerals Engineering*, Vol. 6, No. 6, pp. 607-617, ISSN 0892-6875

Zaghloul, M. R. (2008). A simple theoretical approach to calculate the electrical conductivity of nonideal copper plasma. *Physics of Plasmas,* Vol. 15, No. 4, pp. 701-705, ISSN 1070-664X

Zhang, W., & Yao, Y. L. (2002). Micro scale laser shock processing of metallic components. *Journal of Manufacturing Science and Engineering Transactions of ASME,* Vol. 124, No., pp. 369-378, ISSN 1087-1357

9

Evolution of Calciocarbonatite Magma: Evidence from the Sövite and Alvikite Association in the Amba Dongar Complex, India

S. G. Viladkar

Carbonatite Research Centre, Amba Dongar, Kadipani, District Vadodara
India

1. Introduction

CaCO3-rich carbonatites seem to be the most common and abundant carbonatite in many complexes. The Amba Dongar complex too, is dominantly composed of sövite. Its dike equivalent-(alvikite I and II), though volumetrically less abundant, occur in association with sövite. Majority of sövites samples are composed essentially of CaCO3 and MgO content is low (0.99% average) and goes up to maximum 3% in phlogopite-sövite. A small fraction of periclase was reported earlier from some sövite samples by Viladkar and Wimmenauer, [4]. In this paper I propose that sövites are crystallized from calciocarbonatite magma. Trace elements and REE distribution patterns suggest that sövite fractionated to Alvikite I and II. The C and O isotope data provides evidence to supports the mantle derivation of the carbonatite magma at Amba Dongar.

2. General geology

Amba Dongar carbonatite-alkalic rock diatreme is located on the western periphery of the Deccan flood basalt province in Gujarat State, Western India. This sub-volcanic diatreme consisting of different phases of carbonatites, nephelinites and phonolites, has intruded into the Bagh sandstone and overlying Deccan basalts sequence and has been dated at 65.5 Ma [1].

The outstanding geological feature of the Amba Dongar complex is the bold exposures and clear intrusive relationship between different carbonatite units. Sövite is the predominant type of carbonatite and forms a ring dike around the inner rim of carbonatite breccia (Fig.1). Apart from major ring dike, sövite also forms large and small plugs within the Bagh sandstone terrain. Ankeritic carbonatite intrudes sövite at several places as dikes and large oval plugs while, last in the differentiation sequence, sideritic carbonatite phase intrudes the ankeritic carbonatite as thin dikes, in the southwestern part of the ring structure. Alvikite (phase I), dikes equivalents of sövite, are widespread within as well as outside the ring structure. Alvikite phase II intrude sövite at some places along the ring dike. Nephelinite plugs and phonolite dikes occur in the outer periphery of the ring structure. Large reserves of hydrothermal fluorite deposits (11.5 million tones) are associated with the carbonatites. A vertical zonation of fenitization developed with strongly sodic and sodi-potassic fenites at deeper levels and potassic fenites at higher levels [2] [3]. The central part of the ring structure is occupied by intrusive basalt which replaced the original plug of carbonatite breccia. This youngest intrusive event seems to have occurred at 41.7 Ma [3].

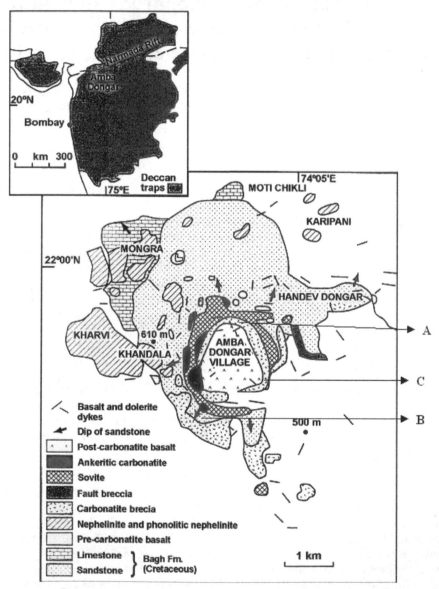

Fig. 1. Geological map of the Amba Dongar carbonatite diatreme with locations of Alvikite centres within sövite ring and outside the ring. A and C are in carbonatite breccia, B in Bagh sandstone.

It has been possible to establish the different phases of sövite intrusion with each phase marked by characteristic texture and mineralogy [3] [4]. The texture of sövite varies from coarse grained in the inner part of the ring in contact with carbonatite breccias to very fine grained in the outermost part of the ring in contact with the sandstone. Sövite also reveals excellent flow structure, and at several places large blocks of fenites are caught up within the flowage structure.

Early crystallization (segregations) of magnetite, apatite, pyrochlore and Nb-zirconolite occur as bands within sövite (Fig. 2). Silico-sövites (phlogopite-sövite and pyroxene-sövite) are not exposed at present level of erosion, however; many xenoliths of these rocks were encountered at deeper levels during mining operations and also in a tunnel driven through the sövite exposure.

Fig. 2. Banding (shown by an arrow) in sövite.

The alvikites vary in thickness from several centimeters to almost a meter. Detailed field investigation of numerous dikes suggests that there are at least two distinct intrusive phases; one either predating or contemporaneous with the main phase of sövite intrusion while the second (alvikite II) clearly postdates the sövite intrusion. In general, alvikites do not show any particular trend in their disposition and, they are located all over the central ring structure as well as outside the ring structure in sandstones and basalts. Three main centres of alvikite (Fig. 1) have been located in and around the ring dike of sövite. Alvikites of phase I intrude mainly carbonatite breccia (Fig. 3), sandstone and basalts while those of the second phase intrude sövite and fenite outcrops. Apart from these exposures alvikites were also exposed in fluorite mine workings and along road cuttings.

Fig. 3. Alvikite I (shown by an arrow) in carbonatite breccia.

The majority of phase I alvikites show thin chilled contacts with the host rocks. In some dikes grain size variation from margin (smaller grains) to center (larger grains) is observed. However, the most conspicuous feature observed around alvikite dikes of this phase (whenever they intrude sandstone) is the feldspathization of country rock sandstone. The fenitization effects along alvikite-basalt contact are not so conspicuous except for a small increase in K (formation of K-feldspar), modal apatite and calcite in the host basalt.

Fig. 4. Alvikite II (shown by an arrow) in sövite.

Second generation alvikites (phase II) intrude sövite (Fig. 4) in the ring dike and some are located in exposures of fenites. Mineral zonation is clearly developed in some of these alvikites with margins exclusively rich in calcite while concentration of magnetite and pyrochlore in the central part of the dikes.

In Amba Dongar, along with intrusive phase of calciocarbonatite, the extrusive phase also occurs in the form of tuff (Fig.5) in the western part at Mongra [20].

Fig. 5. Carbonatite tuff beds (shown by an arrow) in Mongra, West of Amba Dongar.

3. Petrography and mineralogy

Sövite shows various textures such as coarse grained, medium grained and, fine grained porphyritic [6]. In porphyritic types, the euherdal calcite phenocrysts are embedded in fine grained groundmass calcite of a later generation. In silico-sövites primary phlogopite is the most abundant silicate mineral while diopsidic augite is next in abundance. Phlogopite occurs as euherdal to sub-hedral crystals which locally show resorption by surrounding calcite. Occasionally phlogopite forms thin bands with individual crystals showing parallel orientation. The majority of phlogopite crystals are homogenous and do not show any zoning. Accessory minerals in order of decreasing abundance are apatite, magnetite, fluorite, barite, galena, pyrochlore, niobian-zirconolite, zircon, bastnaesite and monazite.

Alvikites, on the other hand, are medium to fine grained equigranular with anhedral grains of calcite and ankeritic calcite (in alvikite II) showing interlocking texture. Porphyritic texture is not uncommon. Some dikes show good flow texture with platy calcite crystals aligned along the direction of flow. All alvikite dikes of phase I have mineralogy broadly similar to coarse sövites whereas the post-sövite alvikites (phase II) are characterized by abundance of phlogopitic-biotite displaying strong zoning with pale brown core and dark brown rims. Oscillatory zoning is not uncommon. Dark brown to reddish brown tetraferriphlogopite occurs as small isolated flakes as well as rims of some strongly zoned crystals [7]. Octahedral magnetite is very common; sometimes it also forms a thin rim around pyrochlore.

Micas from sövites and alvikites were investigated in detail earlier and the details of their evolutionary history from sövite to alvikite are discussed in earlier publication [7]. The chemical differences shown by the two types of micas (Mg-rich in sövite and Fe-rich in alvikite II) and their evolution trends from sövite to Alvikite II are thought to reflect the chemical evolution of the calciocarbonatite magma from which they crystallized.

4. Analytical method

Major and trace elements were determined by X-ray fluorescence using a Philips PW 1450/20 instrument at the Mineralogische-Petrographisches Institute of the Albert-Ludwigs University, Freiburg im Br. Germany. Rare earth elements were determined by the Neutron Activation method at the B.A.R.C., Trombay, Mumbai (Note: where only Ce, La and Nd values mentioned in the analytical tables, these are the XRF analyses. Other REE analyses (La-Lu) are by the Neutron Activation method).

C and O isotopes were determined at the Biogeochemistry section of the Max-Planck Institute, Mainz, Germany using following method

Finely powdered (200 mesh-size) carbonatite samples were reacted for 72 hours with anhydrous (100%) phosphoric acid at 25^0 C [8] [9] to ensure quantitative reaction of dolomite and ankerite components. Carbon and oxygen isotope ratios were determined for CO_2 on a VG PRISM mass spectrometer equipped with a three-collector system allowing the simultaneous collection of the masses 44, 45, and 46. Results were corrected for ^{17}O after Craig [9] and are reported as $\delta^{13}C$ and $\delta^{18}O$ values relative to the Pee Dee Belemnite (PDB) and Standard Mean Ocean Water (SMOW) standards, respectively, with

$$\delta^{13}C = \{[(^{13}C / ^{12}C)sa / (^{13}C / ^{12}C)st] - 1\} * 1000(\text{‰})$$

and

$$\delta^{18}O = \{[(^{18}O / ^{16}O)sa / (^{18}O / ^{16}O)st] - 1\} * 1000(\text{‰})$$

were sa = sample and st = standard. Standard deviations of the measurements were usually smaller than ± 0.1 ‰.

5. Geochemistry

The present section is aimed mainly to show how the parent calciocarbonatite magma at Amba Dongar fractionated. Geochemical differences between sövite and alvikite (phases I and II) during their evolution are brought out using spidergam for trace elements and the chondrite normalized REE patterns. Representative major, trace and rare earth elements anlyses of sövite, alvikite (phase I) and alvikite (phase II) are given in tables 1, 2 and 3 respectively.

Evolution of Calciocarbonatite Magma: Evidence from the Sövite and Alvikite Association in the Amba Dongar Complex, India

199

Sample	1201SOV	1244SOV	1250SOV	1270SOV	187SOV	195SOV	Ave.	SD	CV%
SiO_2	2.33	7.19	0.68	3.82	4.82	2.62	2.97	3.167	0.949
TiO_2	0.96	0.15	0	0	0.41	0	0.09	0.182	1.889
Al_2O_3	0.26	0.18	0	0.17	0.58	0.18	0.16	0.187	1.232
Fe_2O_3	2.22	2.08	1.66	1.1	5.25	1.74	2.63	2.468	0.955
MnO	0.59	0.53	0.73	0.44	0.15	0.66	0.5	0.429	0.826
MgO	0.26	0.16	3.08	0.16	2.05	1.16	0.99	0.972	1.067
CaO	50	47.84	50.57	51.9	47.19	50	49.34	3.511	0.071
Na_2O	0	0.15	0	0.12	0	0	0.04	0.066	1.516
K_2O	0	0	0	0	0.16	0	0.06	0.066	1.232
P_2O_5	0.6	3.47	0.45	1.04	3.99	0.19	1.69	1.469	0.855
CO_2	41.89	36.3	41.52	40.2	33.32	41.74	39.76	2.814	0.071
Total	99.31	98.05	98.69	98.95	97.92	98.29	92.95		
Ba	770	3860	6580	920	980	4130	4080	3125	0.786
Sr	6100	3630	2560	5515	6310	2260	6871	6402	1.018
Nb	1070	250	190	110	80	490	535.88	1179	0.636
Zr	80	110	70	35	260	80	101.51	151.58	2.204
Y	50	230	60	130	80	110	124	77.64	1.427
Th	120	130	60	30	80	40	59	56.64	0.796
U	10	45	0	50	30	0	13.97	30.12	1.988
V	10	0	65		115	50	110.14	157	1.483
La	290	1110	662	1830	666	1080	1040.44	894.4	0.903
Ce	879	2460	1660	2980	1350	2800	2581.14	2697	1.117
Nd	233	550	375	410	446	695	572.05	459	0.862
Sm	34	125	47	44	58	177			
Eu	9.23	19.9	11.3	10.4	14.3	25.3			
Tb	3.09	8.61	3.59	4.32	4.15	7.95			
Yb	4.28	16	3.5	12.7	4.2	8.83			
Lu	0.55	2.33	0.4	1.78	0.55	0.87			

Table 1. Representative analyses of sövite from different parts of ring dike and plugs.

	1	2	3	4	5	6	
Sample	1272AD(I)	1292AD(I)	79AD(I)	86/A/48(I)	BH/15/ALV	V/alk/68	Ave
SiO2	1.64	2.48	1.44	3.78	1.56	6.22	3.65
TiO2	0.00	0.00	0.11	0.02	0.11	0.13	0.09
Al2O3	0.37	0.00	0.11	0.21	0.07	1.78	0.75
Fe2O3	1.20	0.40	2.11	1.03	3.19	1.29	1.84
MnO	0.44	0.20	0.85	0.28	0.16	0.37	0.28
MgO	0.17	0.80	0.24	0.60	0.42	1.46	0.75
CaO	51.68	53.26	50.10	51.65	50.49	48.52	49.42
Na2O	0.20	0.10	0.18	0.06	0.05	0.54	0.16
K2O	0.00	0.00	0.00	0.10	0.00	0.59	0.37
P2O5	0.30	0.00	0.47	0.15	3.24	0.09	0.87
CO2	42.04	42.10	40.78	41.50	39.80	38.90	40.48
Total	98.54	97.34	96.39	99.38	100.49	99.91	
Ba	2815.00	*	9850.00	211.00	485.00	210.00	2960.3
Sr	16410.00	*	5210.00	2407.00	4295.00	1282.00	6610
Nb	40.00	*	160.00	70.00	135.00	10.00	87.9
Zr	20.00	*	80.00	44.00	220.00	78.00	161.5
Y	175.00	*	155.00	70.00	40.00	33.00	226.8
Th	120.00	*	130.00	10.00	40.00	0.00	59.4
U	20.00	*	20.00	0.00	10.00	0.00	62.44
V	*	*	*	33.00	85.00	40.00	23.3
La	1070.00	354.00	1830.00	823.00	310.00	60.00	634.18
Ce	2140.00	680.00	3340.00	1430.00	690.00	120.00	1237.455
Nd	445.00	181.00	612.00	230.00	200.00	45.00	265
Sm	*	30.00	73.00	64.00	*	14.00	
Eu	16.30	7.51	16.20	7.93	*	4.00	
Tb	6.57	2.93	6.65	2.10	*	2.00	
Yb	11.80	6.37	8.60	1.70	*	1.50	
Lu	1.41	0.79	1.22	0.34	*	2.00	

Table 2. Representative analyses of alvikite I.

Sample	1 1246/AD/II	2 1268(II)	3 1269AD(II)*	4 1288(II)	Ave
SiO2	8	4.77	5.19	1	4.11
TiO2	0.1	0.15	0.12	0.23	0.17
Al2O3	0	0.05	0.04	0.04	0.07
Fe2O3	8	10.21	7.9	9.3	8.19
MnO	1.5	0.74	1.43	0.5	0.84
MgO	0.6	0.41	0.45	0.8	0.5
CaO	45	40.2	44.48	46.3	45.08
Na2O	0.1	0.12	0.15	0	0.07
K2O	0.2	0	0	0	0.03
P2O5	2.1	1.12	0.49	1.6	1.01
CO2	37.6	40.5	38.2	39	38.43
Total	100.2	98.27	98.45	98.77	98.26
Ba	12800	10810	8730	8200	8438
Sr	17510	3910	3210	6570	6059
Nb	1600	1890	580	1670	956
Zr	*	140	80	200	91
Y	180	120	130	80	112
Th	40	50	20	110	99
U	*	25	40	5	13
V	1055	200	135	0	269
La	2070	2140	2360	770	1508
Ce	8370	6010	5870	2250	4051
Nd	935	1050	1310	460	826
Sm	*	*	216	*	
Eu	*	*	34.6	*	
Tb	*	*	9.4	*	
Yb	*	*	3.5	*	
Lu	*	*	0.83	*	

Table 3. Representative analyses of Alvikite II.

On the CaO-MgO-Fe^t + MnO plot (Fig 6) there is clear separation between sövite and alvikite II (alvikite I plot in the same field as sövites). The alvikite II are richer in Fe on account of presence of ankeritic calcite in them. Most sövites and alvikites in Amba Dongar show MgO content below 1 wt % while those with MgO more than 1 wt% have either phlogopitic mica or periclase in them. Higher Fe_2O_3 contents of one sövite is due to high modal magnetite in this rock while the high Fe_2O_3 content of phase II alvikites can be attributed to a small amount of oxidized ankeritic calcite present in them.

The fractionation from sövite to alvikites (I and II) can be brought out more clearly on the spidergram for trace elements and chondrite normalized REE patterns. Plotting of average values of trace elements in sövites, alvikite (I) and (II) (Fig.7) show that incompatible trace elements gradually increase from alvikite I → sövite → alvikite (II), and as compared to alvikite I and sövites, alvikite II is enriched in all trace elements except Sr, Zr and Y.

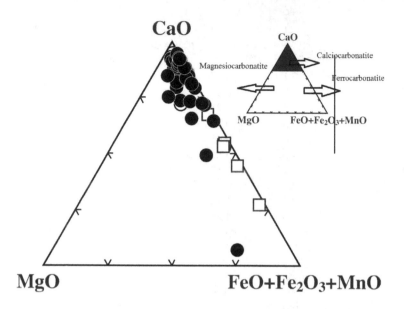

Fig. 6. Amba Dongar sövites (filled circles), Alvikite I (open circles) and Alvikite II (open squares) plotted on CaO-MgO-FeO+Fe_2O_3+MnO.

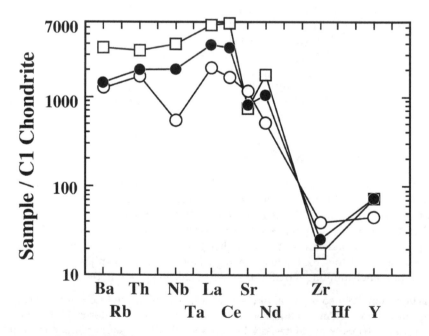

Fig. 7. Spidergram showing trace element variation in averages of Amba Dongar Alvikite I (open circle), Alvikite II (open square) and sövites (filled circle).

Chondrite normalized plots of sövite (two different phases, coarse and fine grained) and alvikites (I and II) of Amba Dongar are shown in Fig. 8. Alvikite II is strongly enriched in all REE in comparison to Alvikite I and sövites. Steep slope of LREE/HREE, similar and parallel REE patterns point to fractionation during evolution of these two types of carbonatites in Amba Dongar.

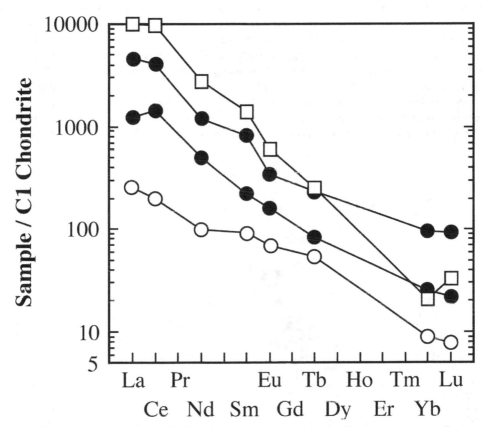

Fig. 8. Chondrite normalized REE distribution patterns of Amba Dongar sövites and alvikites; Open circle — Phase I alvikite, Open square -- Phase II alvikite, Filled circle - 2 phases of sövites (coarse and fine grained).

6. C and O isotopes

Stable isotopes of carbon and oxygen have proved useful in differentiating primary igneous carbonatites from marine carbonates. Relative to the PDB standard, mantle derived carbon has a $\delta^{13}C$ range of –7 to –5 ‰, and relative to the SMOW standard, mantle oxygen has a $\delta^{18}O$ range of +5 to +8 ‰ [10] [11]. However, various types of carbonatites (e.g., calcitic, dolomitic, ankeritic) may have evolved from a set of source values typical of the above range, and have undergone fractional crystallization and subsequent magmatic hydrothermal or meteoric alteration in diverse environments subsequent to their emplacement. Therefore, it is customary to divide carbonatites broadly into two categories, primary and secondary. In the former variety, variations in both the isotope ratios (viz., $\delta^{13}C$ and $\delta^{18}O$) occur mainly due to fractional crystallization; the $\delta^{13}C$ values range from –9 to –1‰, while $\delta^{18}O$ ranges from +5 to +15 ‰ [12] [10]. In the second (altered) variety, the $\delta^{18}O$ values can increase up to +30 ‰, close to that of marine carbonates [13].

The data obtained on C and O isotopes of sövite, alvikites (I & II) supports the observation that calciocarbonatite magma at the Amba Dongar carbonatites complex, is mantle derived as majority of sövite samples and a few alvikites have retained the primary mantle signature. Accordingly, it is seen that the majority of sövite samples and few alvikite I and II plot in the field of mantle derived primary carbonatites in Fig. 9 [14]. Two samples of alvikite I fall within the mantle box while, remaining plot close to the box. Majority of alvikite II plot outside the box and this trend can be explained by process of fractionation. The sövite samples which lie farther away from the mantle box are mostly medium to fine grained sövites. The arrow shows the relation between phenocryst in sövite (star) and the

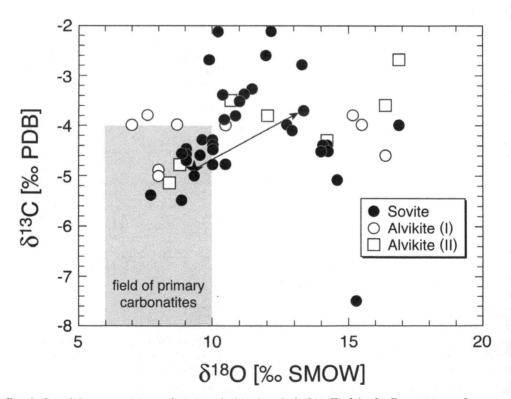

Fig. 9. C and O compositions of sövite, alvikite I and alvikite II of Amba Dongar complex. Open circle — Phase I alvikite, Open square -- Phase II alvikite, Filled circle – 2 phases of sövites (coarse and fine grained).

groundmass sövite. Coarse-grained sövite, in general, has lower $\delta^{13}C$ and $\delta^{18}O$ values compared to fine grained ones, indicating that the fractional crystallization of the carbonatite magma initiated with the crystallization of coarse-grained sövite. As fractionation proceeded, the magma continuously became enriched in water (i.e., the CO_2/H_2O ratio decreased), at which time crystallization of fine-grained sövites started, thus explaining their higher $\delta^{13}C$ and $\delta^{18}O$ values.

It thus can be concluded that the primary carbonatites of Amba Dongar are interpreted to have evolved from a carbonatite parent magma of $\delta^{18}O = 8\%_0$ and $\delta^{13}C = -5.3\%_0$. The initial molar ratio of H_2O to CO_2 that best explains the trend is ~0.9, with a crystallization temperature of ~800 °C. The secondary carbonatites in Amba Dongar appear to be altered mainly by a hydrothermal fluid with H_2O/CO_2 ratio between 0.001 and 0.1 at temperatures between 100 and 200°C. This model also applies to the secondary carbonatites found in carbonatite dikes from Amba Dongar [15].

7. Discussion

There has been an ongoing debate for many years about whether calciocarbonatite magmas exist or whether calcitic carbonatites are cumulate rocks derived from the crystallization of dolomitic carbonatite magma. While this may be true in some cases the present study of Amba Dongar dikes argues very strongly for the existence of a calcitic carbonatite magma (calciocarbonatite) magma, although somewhat magnesian in its early stages.

In Amba Dongar field relations between sövite and two phases of alvikites and petrographic observations such as tabular plates of calcite aligned along flowage, porphyritic texture with phenocrysts of calcite provide convincing evidence to assume that these carbonatites are crystallized from CaCO3-rich melt. The surface exposures of different generations of carbonatites are very clear and injection of alvikite II into sövite provides indisputable evidence of their origin that they being younger than the main sövite intrusion. With the help of geochemistry of major, trace and rare earth elements the fractionation of calciocarbonatite magma from sövite to two generations alvikite is well documented in the earlier section.

Early experimental studies of Wyllie and Tuttle [16] showed that calcite can crystallize as a liquidus phase at a temperature around 650° C at 0.1 GPa pressure.

According to Gittins [17] and Bailey [18] carbonatites can be precipitated from primary calciocarbonatite magma. Bailey listed 8 examples effusive calciocarbonatite [18], (Table 2, p. 6) and majority of them erupted from mantle as primary melt. Mariano and Roedder [19] studied products of Kerimasi carbonatite volcano and concluded that calcite was a primary phase in some flows. It is worth mentioning here that in Amba Dongar too an extrusive phase of calciocarbonatite magma occurs in the form of well bedded tuff and C and O isotopes of an unaltered tuff plots in field of mantle derived primary carbonatites [20]. Jago and Gittins [21] and Gittins and Jago [22] suggest, from experimental evidence, that a small amount of fluorine in carbonatite magma can reduce the temperature of crystallization of calcite which allows crystallization of calcite at atmospheric pressure and conclude that the most of the calcite in extrusive carbonatite is primary magmatic. Dalton and Wood [23] demonstrated experimentally the formation of primary calciocarbonatite magma at low pressure. This is further supported by experimental results of Wyllie and Lee [24]. Additional evidence of existence of primary calciocarbonatite comes from occurrence of globules of calcite in mantle xenoliths which has been considered as evidence for the formation of immiscible $CaCO_3$-rich melts in the mantle [25].

It thus appears that calciocarbonatite magma was the primary magma in Amba Dongar and during its evolution differentiated to alvikite I and II. That such calciocarbonatite magma was initially more magnesian is evident from the presence of phlogopite-sövite and periclase bearing sövite [4].

8. Acknowledgements

I am grateful to John Gittins for discussions, comments and suggestions on the manuscript. I thank: R. Ramesh of PRL, Ahmedabad for discussions on stable isotope data, Mrs Schegel formerly at Mineralogisches-petrographisches Institute, Freiburg, Germany for technical assistance. Dr. P.B. Pawaskar, formerly at BARC, Trombay, Mumbai for the REE analystical work. Part of the analytical work was done during the tenure of an Alexander von Humboldt fellowship in 1980 for which I thank the AvH Foundation, Germany. For the stable isotope determination I thank Mr. Georg Josten and Hedi Oberhansli for help during the laboratory work and Professor Schidlowski for guidance and discussions. I also gratefully acknowledge receipt of a Max-Planck Fellowship for conducting this investigation.

9. References

[1] Ray, J.S., Kanchan Pande and Venkatesan, T.R. (2000) Emplacement of Amba Dongar carbonatite alkalic complex at Cretaceous/Tertiary boundary: Evidence from $^{40}Ar/^{39}Ar$ chronology. Proc. Ind. Aca. Sci. (Earth Planet.Sci) 109 : 39- 47.
[2] Viladkar, S.G. (1986) Fenitization at the Amba Dongar carbonatite alkalic complex. Proce. Sympo. NEMIRAM, Czechoslovakia: pp170-189
[3] Viladkar, S.G. (1996) Geology of the Carbonatite-Alkalic Diatreme of Amba Dongar Gujarat. GMDC Sci. and Research Centre, Ahmedabad : pp1-74
[4] Viladkar, S.G. and Wimmenauer, W. (1992). Geochemical and Petrological studies on the Amba Dongar carbonatites (Gujarat, India). Chem. Erde. 52:277-291.
[5] Viladkar, S.G., Ramesh, R., Avasia, R.K. and Pawaskar, P.B. (2005). Extrusive phase of Carbonatite-Alkalic activity in Amba Dongar complex, Chhota Udaipur, Gujarat. Geol. Soc. India. 66: 273-276.
[6] Viladkar, S.G., (1981): The carbonatites of Amba Dongar, Gujarat, India. Bull. Geol. Soc. Finland. 53: 17-28
[7] Viladkar, S. G. (2000) Phlogopite as indicator of magmatic differentiation in the Amba Dongar carbonatite, Gujarat, India. N. Jb.Miner. Mh. 7: 302-314.

[8] McCrea, J. M. (1950) The isotopic chemistry of carbonates and a palaeotemperature scale. J. Chem. Phys. 18: 849-857

[9] Craig, H., (1957) Isotopic standards for carbon and oxygen and correction factors for mass spectrometric analysis of carbon dioxide. Geochim. Cosmochim. Acta. 12: 133-149.

[10] Deines, P. (1989) Stable isotope variations in carbonatites, In: Carbonatites: genesis and evolution editor K. Bell. Unwin Hyman, London pp301-359.

[11] Keller, J. and Hoefs, J. (1995) Stable isotope characteristics of recent natrocarbonatite from Oldoinyo Lengai, In: Carbonatite Volcanism: Oldoino Lengai and the Petrogenesis of Natrocarbonatites, Editors K. Bell and J. Keller, IAVCE I, *Proc.* Volcanol. 4: pp113-123.

[12] Plyusnin, G. S., Samoylov, V.S., and Gol'shev, S.I.(1980) The $\delta^{13}C$, $\delta^{18}O$ isotope pair method and temperature facies of carbonatites, Doklady Akademi Nank USSR, Seriya Geologiya. 254:1241-1245.

[13] Ray, J. S. and Ramesh, R., 1999. Evolution of carbonatite complexes of Deccan Flood Basalt province: stable carbon and oxygen isotopic constraints *J. Geophys. Res.*B12, 104, 29471-29483.

[14] Taylor, H.P. Jr. Frenchen, J. and E.T. Degens (1967) Oxygen and carbon isotope studies of carbonatites from the Laacher See district, West Germany and the Alnö district, Sweden. Geochim Cosmochim Acta. 31: 407-430

[15] Viladkar, S.G. and R. Ramesh Stable Isotope geochemistry of some Indian carbonatites: Implications for post-emplacement hydrothermal alteration processes (in preparation)

[16] Wyllie, P.J. and O.F. Tuttle (1960a) The system $CaO-CO_2-H_2O$ and the origin of Carbonatites. J. Petrol. 1: 1-96.

[17] Gittins, J. (1989) The origin and evolution of carbonatite magmas. In Carbonatites: Genesis and evolution Editor K. Bell. London, Unwin Hyman : pp580-600.

[18] Bailey, D.K. (1993) Carbonatite magmas. J. Geol. Soc. 150 : 637-651.

[19] Mariano, A. N. and P.L. Roedder (1977) Kermasi: a neglected carbonatite volcano. J. Geol, 91: 449-455.

[20] Viladkar, S.G., Ramesh, R., Avasia, R.K. and Pawaskar, P.B. (2005). Extrusive phase of Carbonatite-Alkalic activity in Amba Dongar complex, Chhota Udaipur, Gujarat. Geol. Soc. India. 66: 273-276.

[21] Jago, B.C. and Gittins, I. (1991) The role of fluorine in carbonatite magma evolution. Nature. 349: 56-58.

[22] Gittins. J and B.C. Jago (1991) Extrusive carbonatites: their origins reappraised in the light of new experimental data. Geol Mag. 128, 4: 301-305.

[23] Dalton J.A. and Wood B.J. (1993) The compositions of primary carbonate melts and their evolution through wall rock reactions in the mantle. Earth & Plan. Sci Lett 119: 511-525.

[24] Wyllie, P. J. and Lee, W.J. (1998) Model system controls on conditions for formation of magnesiocarbonatite and calciocarbonatite magmas from the mantle. J. petrol. 39 : 1885-1893.

[25] Kogarko, L.N., C.N.B. Henderson and H. Pacheko (1995) Primary ca-rich carbonatite Magma and carbonate-silicate-sulphide liquid immiscibility in the upper mantle. Contr. Miner. Petrol. 121: 267-274.

10

Characteristics of Baseline and Analysis of Pollution on the Heavy Metals in Surficial Soil of Guiyang

Ji Wang[1,2] and Yixiu Zhang[1,2]
[1]School of Geographical and Environment Sciences
Guizhou Normal University, Guiyang, Guizhou
[2]Key Laboratory of Remote Sensing Applications in Resources and Environments, Guizhou
China

1. Introduction

The term "Environmental Geochemical Baseline (EGB)" first appeared in the International Geochemical Mapping Program (IGCP259) and the International Geochemical Baseline Program (IGCP360) of International Geo-graph Contrast Program. The definition of EGB refers to natural changes in the concentrations of chemical materials (chemical elements) in the Earth's surface material (Salminen &Tarvainen, 1997). But the definition is becoming clearer with deepening research on EGB. The geochemical baseline reflects the natural concentration of one element in a particular material (e.g. soil, sediment, and rock). At the same time, it can be described as the unitary limit to distinguish the geochemistry backgrounds and anomalies (Salminen & Gregorau-skiene, 2000).

As for the EGB, it is required to establish the archives of the current Earth's surface environment and provide the database to monitor environmental variation. The aim of EGB is to reveal natural changes in mineral and chemical elements so as to make comparisons with anthropogenic influences. The EGB provides the definition of geochemical variation in natural space. It can not only guider policy-makers to make policies toward environmental problems, but also can educate the public who are interested in environmental problems (Darnley, 1997).

All countries attach great importance to the study of EGB, e.g. Mapping of EGB in Europe (Darnley, 1997). For coping with the world EGB studies, China kicked off the program of "Chinese Environment Geochemistry Supervision and Control Network, and the National Dynamic Mapping of Geochemistry Items" in 1992(Chen Hangxin et al., 1998). Use of the EGB to study the environment impact of mining and smelting activities was carried out in the region of Panzhihua, Sichuan Province, Southwest China (Teng Yanguo et al., 2002, 2003).

In this chapter there has been established the surficial soil EGB of heavy metals(Hg, Cd, As, Pb, Cr, Cu, Ni and Zn) in Guiyang City(covering an area of 8046 km²), Guizhou Province. With soil environmental geochemistry research as the main line the spatial distribution of the heavy metals in surficial soil is combined with research on environmentally geochemical mechanism. An appropriate guideline is chosen to distinguish the influence of natural processes from that of anthropogenic processes on soil environment.

2. Materials and methods

2.1 Study area

Guiyang City was selected as the study area which is the capital of Guizhou Province in Southwest China. Guiyang City, situated between east longitude 106°07' to 107°17', north latitude 26°11' to 27°22', lies in the middle of Guizhou Province and on the eastern slope of Yunnan-Guizhou Plateau (Fig.1). Guiyang, with abundant natural resources, ample energy resources and good natural environment, has a mild-moist subtropical climate because of diversity in geographical and topomorphic features, high elevation and low latitude. The total area of Guiyang was 8046 km², including farmland (35.91%), woodland (33.09%), grassland (3.59%), water area (1.89%), construction land (e.g. residential area, industry, mining, transportation, et al.) (6.00%), garden area (0.70%) and no-use land (18.82%)(e.g. wilderness, ribbing, lake-beach, et al.) according to the land use. Soils in the study area mainly include yellow soil (3.335×10⁵ hm², 41.53%), umber soil (1.49%), limestone soil (2.021×10⁵hm², 25.17%), rocky soil (2.17%), coarse-bone soil (12.64%), purple soil (1.93%), marsh soil (0.11%), paddy soil (1.156×10⁵hm², 14.39%), and mountainous meadow soil (0.5%) (Fig.2).

Fig. 1. Map showing the study site in Guizhou.

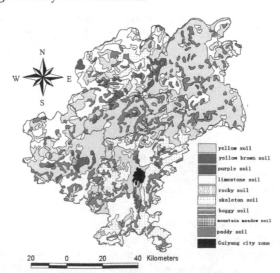

Fig. 2. The distribution of surficial soil of Guiyang, Guizhou.

2.2 Sampling

The snake-form distribution sampling method was adopted because of the bigger sampling area, relief topography and un-uniform soil. The topsoil layer(5~15 cm) was sampled after cover rock and remained roots were removed(HJ/T 166-2004). Guiyang has 1286 villages 83 towns and the sample of Hg, Cd, Pb, Cr and As localities were distributed in 487villages and 75 towns of Guiyang. So the samples account for 37.87% and 90.36%, respectively. The sample of Cu, Ni and Zn localities were distributed in 332 villages and 50 towns, and the samples account for 25.82% and 60.24%. Hg, Cd, Pb, Cr and As localities are shown in Fig. 3. Cu, Ni and Zn localities are shown in Fig. 4.

The soil samples were collected 67 at January19 to March 4, and 420 at July 11 to October 11.

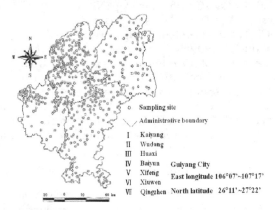

Fig. 3. The distribution on of Cd, Hg, Cr, Pb, As sites in Guiyang, Guizhou.

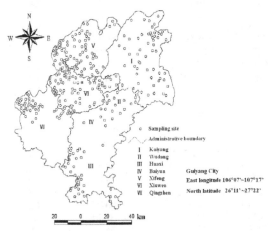

Fig. 4. The distribution on of Cu, Zn, Ni sites in Guiyang, Guizhou.

2.3 Analytical techniques

The content of As was digested with a mixture acids: H_2SO_4-HNO_3-$HClO_4$ (H_2SO_4 / HNO_3/ $HClO_4$, 1:1:3) and using diethyl disulfide generation amino acid silver spectrophotometric method to determine the contents of As in the samples(GB/T17134-1997). The limits of determination were 0.5 mg/kg (As) according to 0.5g sample which was dispelled in 50ml. The content of Pb and Cd were digested with mixture acids: HCl-HNO_3- HF- $HClO_4$, and using graphite furnace atomic absorption spectrometry to determine the contents of Pb and Cd, the limits of determination were 0.1mg/kg (Pb), 0.01mg/kg (Cd) according to 0.5g samples which were dispelled in 50ml(GB/T17141—1997). The total content of Cr were determined by diethyl carbon phenol by two spectrophotometric method after the samples were digested with a mixture acids: HCl-HNO_3-HF, and the limits of determination was 1.0mg/kg(Cr) according to 0.5g samples which was dispelled in 50ml (GB/T17137—1997). Using flame-atomic absorption spectrophotometry to determine the contents of Cu, Zn and Ni in the samples(Cu and Zn: GB/T17138—1997; Ni: GB/T17139—1997). The limits of determination were 1.0 mg/kg (Cu), 5.0 mg/kg (Ni), 0.5 mg/kg (Zn) according to 0.5g sample which was dispelled in 50ml. The samples of Cu, Ni and Zn were digested with HCl-HNO_3- HF- $HClO_4$. The total content of Hg was digested with a mixture of ultrapure acids: H_2SO_4-HNO_3-$KMnO_4$(H_2SO_4/HNO_3,1:1) and analyzed by cold atomic absorbent spectrophotometry, the limits of determination was 0.005mg/kg(Hg) according to 2g samples which was dispelled in 50ml(GB/T17136—1997). The concentrations of Hg, Pb, Cd, Cr, Cu, As, Ni and Zn in the solution were measured under the optimum condition. For quality assurance and quality control, reagent blanks, 20% duplicated samples and sol standard reference materials GSS-1, GSS-3, GSS-4 obtained from Center of National Standard Reference Material of China were prepared and analyzed with the same procedure and reagents.

The table 1 showed that the accuracy and precision of testing of the above.

The available data sets were analyzed using the SPSS 16.0, ArcGIS, and ArcView.

Elements	SU	SS	GV(mg/kg)	ORM(mg/kg)	IRSD (%)	RSDBR(%)	RE(%)
	14	GSS-1	10.7±0.8	10.7	2.0	5.6	0.0
As	15	GSS-3	15.9±1.3	17.1	1.3	4.3	7.5
	12	GSS-4	11.4±0.7	11.4	3.8	4.8	0.0
Cd	25	GSS-1	0.083±0.011	0.080	3.6	6.2	-3.6
	28	GSS-3	0.044±0.014	0.045	4.1	8.4	2.3
Cr	16	GSS-1	57.2±4.2	56.1	2.0	9.8	-1.9
	18	GSS-3	98.0±7.1	93.2	2.3	8.3	-4.9
	35	GSS-1	20.9±0.8	20.7	2.3	6.8	-0.96
Cu	34	GSS-3	29.4±1.6	29.2	2.0	4.8	-0.68
	30	GSS-4	26.3±1.7	25.6	2.3	3.9	-2.7
	25	GSS-1	0.016±0.003	0.016	6.2	32.5	0.0
Hg	26	GSS-3	0.112±0.012	0.100	3.4	20.0	-10.7
	24	GSS-4	0.021±0.004	0.019	8.4	20.5	-9.5
	29	GSS-1	29.6±1.8	29.1	2.5	8.4	-1.7
Ni	32	GSS-3	33.7±2.1	34.0	2.6	6.0	0.89
	33	GSS-4	32.8±1.7	34.1	2.9	9.1	4.0
Pb	19	GSS-1	23.6±1.2	23.7	4.2	7.3	0.42
	21	GSS-3	33.3±1.3	33.7	3.9	8.6	1.2
	32	GSS-1	55.2±3.4	56.2	2.8	7.3	1.8
Zn	31	GSS-3	89.3±4.0	88.4	1.6	5.0	1.0
	31	GSS-4	69.1±3.5	68.1	3.2	4.1	-1.4

SU: sample numbers; SS: standard samples; GV: guaranteed value; ORM: overall mean
IRSD: indoor relative standard deviation; RSDBR: relative standard between room; RE: relative error

Table 1. The accuracy and precision of contents of heavy metals in soil.

3. Experimental results

Table2 indicated that the statistical analysis results of soil heavy metals concentrations in Guiyang city. The mean value, standard deviation and maximum of As separately are 18.09 mg/kg, 11.57 mg/kg and 79.30 mg/kg in the surficial soil in Guiyang. The content of As in 95.9 per cent sample are smaller than 40 mg/kg., of Cd separately are 0.302mg/kg, 0.363mg/kg, 2.620mg/kg and 95.7 percent smaller than 1.000mg/kg., of Cr 75.3 mg/kg, 37.3 mg/kg, 271.0 mg/kg and 95.9 percent smaller than 150.0 mg/kg, of Cu separately are 43.1mg/kg, 30.3mg/kg, 213.0mg/kg and 94.6 percent smaller than 100.0mg/kg, of Pb are 43.2mg/kg, 31.3mg/kg, 318.9mg/kg and 95.5 per cent smaller than 100.0mg/kg, of Hg are 0.222mg/kg, 0.531mg/kg, 7.030mg/kg and 98.2 smaller than 1.000mg/kg, of Ni are 38.3mg/kg, 14.9mg/kg, 102.5mg/kg and 95.8 percent smaller than 70.0mg/kg, of Zn are 84.7mg/kg, 49.8mg/kg, 385.0mg/kg and 94.3 percent smaller than 150.0mg/kg.

Elements	SN	Min(10^{-6})	Max(10^{-6})	Mean(10^{-6})	SD	CV
As	486	2.70	79.00	18.09	11.57	0.64
Cd	487	0.001	2.620	0.302	0.363	1.20
Cr	487	6.9	271.0	75.3	37.3	0.50
Cu	333	2.1	213.0	43.1	30.3	0.70
Pb	487	0.9	318.9	43.2	31.3	0.7
Hg	487	0.010	7.030	0.222	0.531	2.39
Ni	333	9.2	102.5	38.3	14.9	0.39
Zn	333	0.1	385.0	84.7	49.8	0.59

SN: sample numbers; SD: standard deviation; CV: coefficient of variation

Table 2. The statistical analysis results of heavy metals concentration in soil, Guiyang city.

4. Analysis and discussions

4.1 Results analysis

4.1.1 Establishment of the baselines of heavy metals in surficial soil

4.1.1.1 Establishment of the baselines of As in surficial soil

4.1.1.1.1 Relatively accumulative total amount analysis

Assuming the concentrations of chemical elements in natural surficial soil are of logarithmic normal distribution, the inflexion in the figure of relatively accumulative density to the concentration of chemical element represents the boundary line between the background value and the abnormal value. The range of baseline values of chemical elements is the average value plus double standard deviation of less than the boundary value (Lepeltier, 1969).

The double logarithmic figure of relatively accumulative density (RAD) to the concentration of the chemical element As in topsoil of Guiyang was shown in Fig.5. The inflexion (black points in the figure) is 17.2 mg/kg. So the range of baseline values of As in topsoil of Guiyang is 7.75~15.15 mg/kg, i.e., the average value of 11.45mg/kg plus a double standard deviation of 3.70 mg/kg less than 17.2 mg/kg.

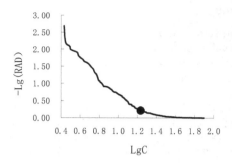

Fig. 5. The logarithm of the concentrations on As and the logarithmic curve of relatively accumulative density in surficial soil.

4.1.1.1.2 Relatively accumulative frequency

The normal decimal coordinates are adopted. There are two inflexions in the figures of relatively accumulative frequency to the concentration of chemical element. The lower one may represent the upper limit of the baseline of chemical elements and the upper one may represent the lower limit of abnormity, i.e., the influence of human activity on the two inflexions. The average or median that is less than the lower inflexion can be regarded as the baseline of chemical elements. The metrical values between the two inflexions may have something with the influence of human activities, or have nothing to do. If the distribution curve looks like a straight line, the measured values may represent the baseline range (Bauer & Bor,1993,1995); Bauer et al.1992; Matschullatetc, 2000)

The figure of relatively accumulative frequency to the concentration of As in topsoil of Guiyang is shown in Fig.6. There are two inflexions: one is 13.0 mg/kg and the other is 29.0mg/kg. So the first inflexion (13.0 mg/kg) represents the upper limit of baseline values of As in topsoil of Guiyang. The average of 9.20 mg/kg or the median of 9.04 mg/kg less than the first-inflexion can be regarded as the baseline of As in topsoil of Guiyang. The second inflexion (29.0 mg/kg) may represent abnormity, i.e., the influence of human activity.

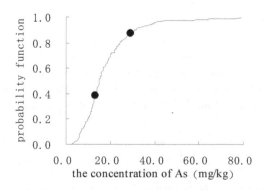

Fig. 6. The probability functions of As in surficial soil.

Comprehensively considering the results of calculation using the two kinds of methods, we respectively take 9.04mg/kg as the baseline values of As in topsoil of Guiyang.

4.1.1.2 Establishment of the baselines of Cd in surficial soil

4.1.1.2.1 Relatively accumulative total amount analysis

The double logarithmic figure of RAD to the concentration of the chemical element Cd in topsoil of Guiyang was shown in Fig.7. The inflexion (black points in the figure) is 0.189 mg/kg. So the range of baseline values of Cd in topsoil of Guiyang is 0.029~0.123 mg/kg, i.e. the average value of 0.076mg/kg pluses double standard deviation of 0.047 mg/kg less than 0.189mg/kg.

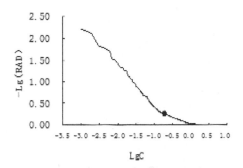

Fig. 7. The logarithm of the concentration on Cd and the logarithmic curve of relatively accumulative density in surficial soil.

4.1.1.2.2 Relatively accumulative frequency

The figure of relatively accumulative frequency to the concentration of Cd is shown in Fig.8. One inflexion is 0.149 mg/kg and the other is 1.010 mg/kg. So the first inflexion (0.149 mg/kg) represents the upper limit of baseline values of Cd in topsoil of Guiyang. The average of 0.068 mg/kg or the median of 0.068 mg/kg less-than the first inflexion can be regarded as the baseline of Cd in topsoil of Guiyang. The second inflexion (1.010mg/kg) may regard as represent abnormity, i.e., the influence of human activities.

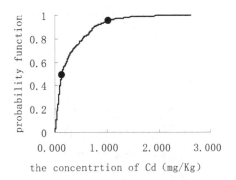

Fig. 8. The probability functions of Cd in surfucial soil of Guiyang.

Comprehensively considering the results of calculation using the two kinds of methods, we respectively take 0.068 mg/kg as the baseline values of Cd in topsoil of Guiyang.

4.1.1.3 Establishment of the baselines of Cu in surficial soil

4.1.1.3.1 Relatively accumulative total amount analysis

The double logarithmic figure of RAD to the concentration of the chemical element Cu in topsoil of Guiyang is shown in Fig.9. The inflexion (black points in the figure) is 32.6 mg/kg. So the range of baseline values of Cu in topsoil of Guiyang is 14.2~28.4 mg/kg, i.e., the average value of 21.3mg/kg pluses a double standard deviation of 7.1 mg/kg less than 32.6 mg/kg.

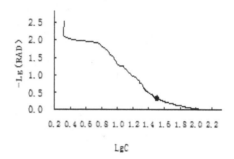

Fig. 9. The logarithm of the concentrations on Cu and the logarithmic curve of relatively accumulative density in surficial soil.

4.1.1.3.2 Relatively accumulative frequency

The figure of relatively accumulative frequency to the concentration of Cu in topsoil of Guiyang is shown in Fig.10. There are two inflexions: one is 28.1 mg/kg and the other is 68.4 mg/kg. So the first inflexion (28.1 mg/kg) represents the upper limit of baseline values of Cu in topsoil of Guiyang. The average of 18.8 mg/kg or the median of 21.9 mg/kg less than the first-inflexion can be regarded as the baseline of Cu in topsoil of Guiyang. The second inflexion (68.4 mg/kg) may represent abnormity, i.e., the influence of human activity.

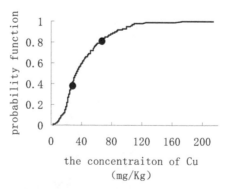

Fig. 10. The probability functions of Cu in surficial soil.

Comprehensively considering the results of calculation using the two kinds of methods, we respectively take 18.8 mg/kg as the baseline values of Cu in topsoil of Guiyang.

4.1.1.4 Establishment of the baselines of Zn in surficial soil

4.1.1.4.1 Relatively accumulative total amount analysis

The double logarithmic figure of RAD to the concentration of the chemical element Zn in topsoil of Guiyang is shown in Fig.11. The inflexion is 114.0 mg/kg. So the range of baseline values of Zn in topsoil of Guiyang is 46.5～91.3 mg/kg, i.e., the average value of 68.9mg/kg pluses double standard deviation of 22.4 mg/kg less than 114.0mg/kg.

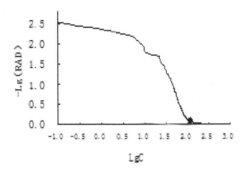

Fig. 11. The logarithm of the concentrations on Zn and the logarithmic curve of relatively accumulative density in surficial soil.

4.1.1.4.2 Relatively accumulative frequency

The figure of relatively accumulative frequency to the concentration of Zn in topsoil of Guiyang is shown in Fig.12. There are two inflexions. One inflexion is 56.5 mg/kg and the other is 112.0 mg/kg. So the first inflexion (56.5 mg/kg) represents the upper limit of baseline values of Zn in topsoil of Guiyang. The average of 41.6 mg/kg or the median of 46.3 mg/kg less-than the first inflexion can be regarded as the baseline of Zn in topsoil of Guiyang. The second inflexion (112.0 mg/kg) may regard as represent abnormity, i.e., the influence of human activities.

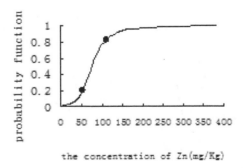

Fig. 12. The probability functions of Zn in surficial soil.

Comprehensively considering the results of calculation using the two kinds of methods, we respectively take 46.3 mg/kg as the baseline values of Zn in topsoil of Guiyang.

4.1.1.5 Establishment of the baselines of Pb in surficial soil

4.1.1.5.1 Relatively accumulative total amount analysis

The double logarithmic figure of RAD to the concentration of the chemical element Pb in topsoil of Guiyang is shown in Fig.13. The inflexion (black points in the figure) is 26.8 mg/kg. So the range of baseline values of Pb in topsoil of Guiyang is 14.0~25.4 mg/kg, i.e., the average value of 19.7mg/kg pluses a double standard deviation of 5.7 mg/kg less than 26.8 mg/kg.

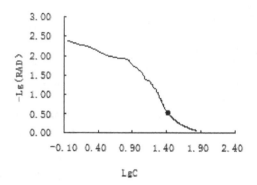

Fig. 13. The logarithm of the concentrations on Pb and the logarithmic curve of relatively accumulative density in surficial soil.

4.1.1.5.2 Relatively accumulative frequency

The figure of relatively accumulative frequency to the concentration of Pb in topsoil of Guiyang is shown in Fig.14. There are two inflexions: one is 20.4 mg/kg and the other is 70.1mg/kg. So the first inflexion (20.4 mg/kg) represents the upper limit of baseline values of Pb in topsoil of Guiyang. The average of 16.0 mg/kg or the median of 14.8 mg/kg less than the first-inflexion can be regarded as the baseline of Pb in topsoil of Guiyang. The second inflexion (70.1mg/kg) may represent abnormity, i.e., the influence of human activity.

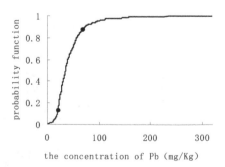

Fig. 14. The probability functions of Pb in surficial soil of Guiyang.

Comprehensively considering the results of calculation using the two kinds of methods, we respectively take 14.8 mg/kg as the baseline values of Pb in topsoil of Guiyang.

4.1.1.6 Establishment of the baselines of Hg in surficial soil

4.1.1.6.1 Relatively accumulative total amount analysis

The double logarithmic figure of RAD to the concentration of the chemical element Hg in topsoil of Guiyang is shown in Fig.15. The inflexion (black points in the figure) is 0.082 mg/kg. So the range of baseline values of Hg in topsoil of Guiyang is 0.031~0.075 mg/kg, i.e., the average value of 0.053mg/kg pluses double standard deviation of 0.022 mg/kg less than 0.082mg/kg.

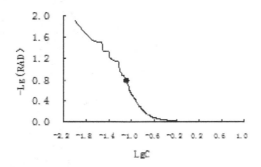

Fig. 15. The logarithm of the concentrations on Hg and the logarithmic curve of relatively accumulative density in surficial soil.

4.1.1.6.2 Relatively accumulative frequency

The figure of relatively accumulative frequency to the concentration of Hg in topsoil of Guiyang is shown in Fig.16. There are two inflexions. One inflexion is 0.072 mg/kg and the other is 0.530 mg/kg. So the first inflexion (0.072mg/kg) represents the upper limit of baseline values of Hg in topsoil of Guiyang. The average of 0.050 mg/kg or the median of 0.045 mg/kg less-than the first inflexion can be regarded as the baseline of Hg in topsoil of Guiyang. The second inflexion (0.530 mg/kg) may regard as represent abnormity, i.e., the influence of human activities.

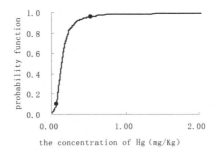

Fig. 16. The probability functions of Hg in surficial soil.

Comprehensively considering the results of calculation using the two kinds of methods, we respectively take 0.045 mg/kg as the baseline values Hg in topsoil of Guiyang.

4.1.1.7 Establishment of the baselines of Cr in surficial soil

4.1.1.7.1 Relatively accumulative total amount analysis

The double logarithmic figure of RAD to the concentration of the chemical element Cr in topsoil of Guiyang is shown in Fig.17. The inflexion (black points in the figure) is 67.8 mg/kg. So the range of baseline values of Cr in topsoil of Guiyang is 31.0~59.8mg/kg, i.e., the average value of 45.4mg/kg pluses a double standard deviation of 14.4 mg/kg less than 67.8 mg/kg.

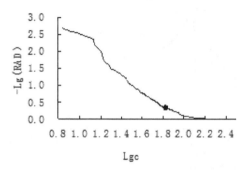

Fig. 17. The logarithm of the concentrations on Cr and the logarithmic curve of relatively accumulative density in surficial soil.

4.1.1.7.2 Relatively accumulative frequency

The figure of relatively accumulative frequency to the concentration of Cr in topsoil of Guiyang is shown in Fig.18. There are two inflexions: one is 63.8 mg/kg and the other is 100.2 mg/kg. So the first inflexion (63.8 mg/kg) represents the upper limit of baseline values of Cr in topsoil of Guiyang. The average of 45.7 mg/kg or the median of 44.0 mg/kg less than the first-inflexion can be regarded as the baseline of Cr in topsoil of Guiyang. The second inflexion (100.2 mg/kg) may represent abnormity, i.e., the influence of human activity.

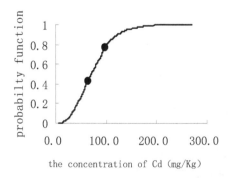

Fig. 18. The probability functions of Cr in surficial soil.

Comprehensively considering the results of calculation using the two kinds of methods, we respectively take 44.0 mg/kg as the baseline values of Cr in topsoil of Guiyang.

4.1.1.8 Establishment of the baselines of Ni in surficial soil

4.1.1.8.1 Relatively accumulative total amount analysis

The double logarithmic figure of RAD to the concentration of the chemical element Ni in topsoil of Guiyang is shown in Fig.19. The inflexion is 27.6 mg/kg. So the range of baseline values of Ni in topsoil of Guiyang is 18.1~26.7 mg/kg, i.e., the average value of 22.4 mg/kg pluses double standard deviation of 4.3 mg/kg less than 27.6mg/kg.

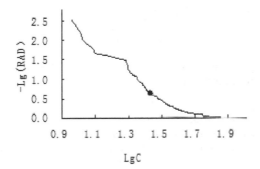

Fig. 19. The logarithm of the concentrations on Ni and the logarithmic curve of relatively accumulative density in surficial soil.

4.1.1.8.2 Relatively accumulative frequency

The figure of relatively accumulative frequency to the concentration of Ni in topsoil of Guiyang is shown in Fig.20. There are two inflexions. One inflexion is 27.8 mg/kg and the other is 57.0 mg/kg. So the first inflexion (27.8 mg/kg) represents the upper limit of baseline values of Ni in topsoil of Guiyang. The average of 19.5 mg/kg or the median of 17.0 mg/kg less-than the first inflexion can be regarded as the baseline of Ni in topsoil of Guiyang. The second inflexion (57.0 mg/kg) may regarded as represent abnormity, i.e., the influence of human activities.

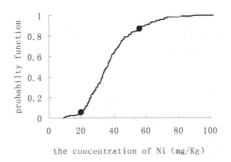

Fig. 20. The probability functions of Ni in surficial soil.

Comprehensively considering the results of calculation using the two kinds of methods, we respectively take 17.0 mg/kg as the baseline values of Ni in topsoil of Guiyang.

4.2 Discussions

4.2.1 Geo-accumulation analysis of heavy metals in surficial soil

Geo-accumulation Index was commonly called Muller Index (Muller, 1969; Forstner & Muller, 1981), was widely used in studying quantitative index for heavy metals pollution in sediments (Forstner et al., 1990; Chen Cuihua et al.,2008; Yi xiu et al.,2010; Hu Mianhao, 2010), and the expression as this.

$$I_{geo} = \log_2 \left[\frac{C_n}{1.5 \bullet BE_n} \right] \qquad (1)$$

C_n represents the concentration of element n in sample. BE_n means the baseline concentration, 1.5 was modified index for characterizing sedimentary characteristics, rocky and other effects.

Geo-accumulation Index can be divided into several levels, e.g. it was divided into seven levels by Forstner (referred to hereafter as F classification), and five levels by Anon (referred to hereafter as A classification). It indicated pollution degrees of heavy metals by different classes of I_{geo}.

F classification			A classification		
I_{geo}	levels	Pollution degrees	I_{geo}	levels	Pollution degrees
<0	1	W.P.	<0	1	W.P. or slight pollution
0~1	2	W.P. to M.P.	0~1	2	M.P.
1~2	3	M.P.	1~3	3	M.P. or S.P.
2~3	4	M.P. to S.P.	3~5	4	S.P.
3~4	5	S.P.	>5	5	S.S.P.
4~5	6	S.P. to S.S.P.			
>5	7	S.S.P.			

W.P.: without pollution; M.P.: mid-pollution; S.P.: strong pollution; S.S.P.: super strong pollution

Table 3. Degrees of pollution by heavy metals indicated by different classes of I_{geo}.

The means of Geo-accumulation indexes of heavy metals in surficial soil of Guiyang city were analyzed (Fig.21 to Fig.28). By the results, we get the surficial soil in 41 per cent of the Guiyang did not suffer the arsenic contaminative, 43 per cent is between without pollution to mid-pollution, 14 per cent mid-pollution, only 2 per cent is between mid-pollution to strong pollution. In 40 per cent did not suffer the cadmium contaminative, 19 per cent between without pollution to mid-pollution, 14 per cent mid-pollution, 19 per cent between mid-pollution to strong pollution, 7 per cent strong pollution, 1 per cent between strong pollution to supper strong pollution. In 46 per cent did not suffer the chromium contaminative, 47 per cent between without pollution to mid-pollution, 6.8 per cent mid-pollution, only 0.2 per cent between mid-pollution to strong pollution. In 38 per cent did not suffer the copper contaminative, 38 per cent between without pollution to mid-pollution, 22 per cent mid-pollution, only 2 per cent between mid-pollution to strong pollution. In 18 per cent did not suffer the lead contaminative, 47 per cent between without pollution to mid-pollution, 28 per cent mid-pollution, only 3 per cent between mid-pollution to strong pollution. In 12 per cent did not suffer the mercury contaminative, 37 per cent between without pollution to mid-pollution, 36 per cent mid-pollution, 11 per cent between mid-pollution to strong pollution, 2 per cent strong pollution, 1 per cent supper between strong pollution to supper strong pollution, 1 per cent supper strong pollution. In 19.2 per cent did not suffer the nickel contaminative, 63.7 per cent between without pollution to mid-pollution, 16.8 per cent mid-pollution, only 0.3 per cent between mid-pollution to strong pollution. In 41 per cent did not suffer the zinc contamination, 50per cent between without pollution to mid-pollution, 7 per cent mid-pollution, only 2 per cent between mid-pollution to strong pollution.

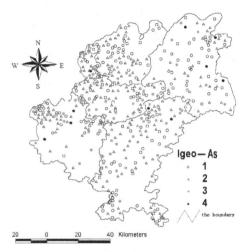

Fig. 21. Distribution of I_{geo} for As in surficial soil.

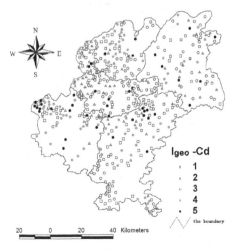

Fig. 22. Distribution of I_{geo} for Cd in surficial soil.

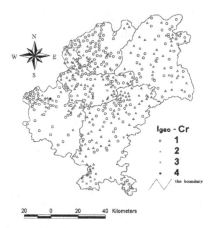

Fig. 23. Distribution of I_{geo} for Cr in surficial soil.

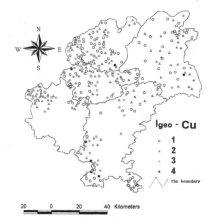

Fig. 24. Distribution of I_{geo} for Cu in surficial soil.

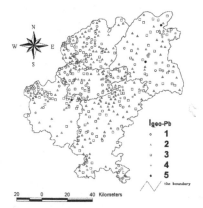

Fig. 25. Distribution of I_{geo} for Pb in surficial soil.

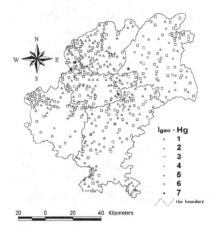

Fig. 26. Distribution of I_{geo} for Hg in surficial soil.

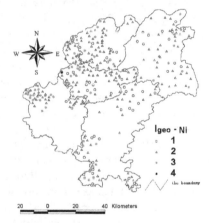

Fig. 27. Distribution of I_{geo} for Ni in surficial soil.

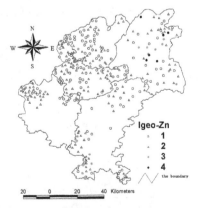

Fig. 28. Distribution of I_{geo} for Zn in surficial soil.

4.2.2 Contamination degree analysis of heavy metals in surficial soil

Contamination Degree (CD) was the most intuitive and commonly used one of the parameters for evaluating heavy metals pollution. The parameter represented the content of heavy metals in soil was over the national standard, and it was expressed as shown.

$$CD = \frac{c_i}{c_A} - 1 \qquad (2)$$

c_i represents the analysis value in i sample of an element (the concentration of an element in sample). c_A means the maximum limit of the element concentration in environment, which was commonly the Quality Standard of Soil Environment. The primary standard of national quality standard in soil environment (GB15618-1995) was used to calculate the heavy metals pollution degree for c_A.

The contamination degrees of heavy metals in surficial soil of Guiyang city were analyzed (Table 4 to Table 11). The results indicated that the maximum of arsenic contaminative degree of surficial soil in Guiyang is 4.27. 50.2 per cent surficial soil did not suffer the pollution. The contaminative degree in 97 per cent surficial soil in Guiyang is smaller than 2 and the total contaminative degree is slightly over zero. So the surficial soil of Guiyang suffers the slight pollution. Of cadmium are 12.1. 57.9 per cent surficial soil did not suffer the pollution. 96 per cent is smaller than 4 and the total is over zero. So suffers pollution of Cd. Of chromium are 2.01. 69 per cent did not suffer the pollution, 30.6per cent slight pollution. The total is less than zero. So not suffer pollution of Cr. Of copper are 5.09. 53.2 per cent did not suffer the pollution. The total is slightly over zero. So suffers slight pollution of Cu. Of lead are 8.11. 49.9 per cent did not suffer the pollution. The total is slightly over zero. So suffers slight pollution of Pb. Of mercury are 45.87. 56.1 per cent did not suffer the pollution. The total is over zero. So suffers pollution of Hg. Of nickel is 1.56. 64 per cent did not suffer the pollution. The total is less than zero. So not suffer pollution of Ni. Of zinc are 2.85. 77.8 per cent did not suffer the pollution. The total is less than zero. So not suffer pollution of Zn. Consideration the pollution to join with 8 kinds of heavy metals, 40.2 per cent have no contamination of heavy metal, 15 per cent from no pollution to slight pollution, 36.1 per cent slightly pollution, 7.2 per cent mid-pollution, 1.4 per cent serious pollution in the surficial soil of Guiyang.

CD_{As}	$x \leq 0$	$0 < x \leq 1$	$1 < x \leq 2$
Number	244	187	42
Ratio	50.2%	38.5%	8.6%
CD_{As}	$2 < x \leq 3$	$3 < x \leq 4$	$x > 4$
Number	6	4	3
Ratio	1.2%	0.8%	0.6%

Note: Average: 0.206; Min:-0.81; Max: 4.27

Table 4. The contamination degree on As in surficial soil of Guiyang city.

CD_{Cd}	$x \leq 0$	$0 < x \leq 1$	$1 < x \leq 2$	$2 < x \leq 3$
Number	282	72	48	42
Ratio	57.9%	14.8%	9.9%	8.6%
CD_{Cd}	$3 < x \leq 4$	$4 < x \leq 5$	$5 < x \leq 6$	$x > 6$
Number	22	7	5	9
Ratio	4.5%	1.4%	1.0%	1.9%
Note: Average:0.51; Min:-0.995; Max:12.1				

Table 5. The contamination degree on Cd in surficial soil of Guiyang city.

CD_{Cr}	≤ 0	$0 < x \leq 0.5$	$0.5 < x \leq 1$
Number	337	118	24
Ratio	69.2%	24.3%	4.9%
CDCr	$1 < x \leq 1.5$	$x > 1.5$	
Number	7	1	
Ratio	1.4%	0.2%	
Note:Average:-0.16;Min:-0.92; Max:2.01			

Table 6. The contamination degree on Cr in surficial soil of Guiyang city.

CD_{Cu}	≤ 0	$0 < x \leq 1$	$1 < x \leq 2$
Number	177	99	43
Ratio	53.2%	29.7%	12.9%
CD_{Cu}	$2 < x \leq 3$	$x > 3$	
Number	11	3	
Ratio	3.3%	0.9%	
Note: Average:0.23; Min:-0.94; Max:5.09			

Table 7. The contamination degree on Cu in surficial soil of Guiyang city.

CD_{Pb}	≤ 0	$0 < x \leq 1$	$1 < x \leq 2$
Number	243	189	37
Ratio	49.90%	38.80%	7.60%
CD_{Pb}	$2 < x \leq 3$	$3 < x \leq 4$	$x > 4$
Number	7	8	3
Ratio	1.50%	1.60%	0.60%
Note:Average:0.23;Min:-0.98; Max:8.11			

Table 8. The contamination degree on Pb in surficial soil of Guiyang city.

CD_{Hg}	≤0	0< x≤1	1< x≤2	2<x≤3
Number	273	155	33	10
Ratio	56.1%	31.8%	6.8%	2.1%
CD_{HG}	3< x≤4	4< x≤5		x>5
Number	6	1		9
Ratio	1.2%	0.2%		1.8%
Note: Average:0.48; Min:-0.93; Max:45.87				

Table 9. The contamination degree onHg in surficial soil of Guiyang city.

CD_{Ni}	≤0	0< x≤0.5	0.5< x≤1
Number	213	88	27
Ratio	64.0%	26.4%	8.1%
CD_{Ni}	1<x≤1.5	x>1.5	
Number	4	1	
Ratio	1.2%	0.3%	
Note: Average :-0.04; Min:-0.77; Max:1.56			

Table 10. The contamination degree on Ni in surficial soil of Guiyang city.

CD_{Zn}	≤0	0< x≤0.5	0.5< x≤1	1<x≤1.5
Number	259	52	11	5
Ratio	77.8%	15.6%	3.3%	1.5%
CD_{Zn}	1.5<x≤2	2<x≤2.5		x>2.5
Number	2	2		2
Ratio	0.6%	0.6%		0.6%
Note: Average:-0.15; Min:-1.00; Max:2.85				

Table 11. The contamination degree on Zn in surficial soil of Guiyang city.

4.2.3 The correlation of heavy metal elements in surficial soil

The correlation of heavy metals were analyzed and twin-elements that correlation coefficient reached extremely significant level($P<0.01$) were As−Pb, Cd−Cr, Ni−Cd, Cu−Cr, Cr−Ni, Cr−Zn, Cu−Ni, Cu−Pb, Cu−Zn, Ni−Zn, Pb−Zn，the twin-elements which correlation was $P < 0.05$ were As−Zn, Cd−Hg, Cd−Zn, Pb−Ni by using SPSS16.0 with CA(Table 12).The correlation between heavy metals mainly because of heavy metal elements between parent rocks associated. For example, the correlation of metalloid element As and Pb reached extremely significant level, of As and Zn reached significant level, for As ore often associated in the sulfide mineral of Pb and Zn, and were produced together with other minerals such as pyrite and sphalerite. The content of As was less than 5-10000mg/kg in the galena, abnormal area were formed when As were released in the ore mining and smelting

(Liao Z J, 1992). Pb-Zn deposits were mainly distributed in the northeast area, northwest area, south-central area and southwest area (Xu H N and Xu J l, 1996). As and its compound were often accompanied in the non-ferrous metal and precious metal ore, arsenide was distributed in the kinds of intermediate product (Li Q and Mo D L, 1997). The correlation of Cd and Zn reached significant, because there was no separate Cd ore, Cd was often accompanied with Zn ore, Cd was generally existed in the Zn ore as the forms of CdS and $CdCO_3$, and the concentration of Cd was between 0.01% and 0.5%.

	As	Cd	Cr	Cu	Hg	Ni	Pb	Zn
As	1						** (+)	* (+)
Cd		1	** (−)		* (+)	** (+)		* (−)
Cr			1	** (+)		** (+)		** (+)
Cu				1		** (+)	** (−)	** (+)
Hg					1			
Ni						1	* (−)	** (+)
Pb							1	** (+)
Zn								1

*: significant level(P<0.05); **: extremely significant level(P<0.01);
(+): positive correlation; (−): negative correlation

Table 12. The relativities between contaminating elements in surficial soil of Guiyang.

4.2.4 The influences of soil types in heavy metals pollution

On the basis of measured values in the different characteristics of soil types on heavy metals in surficial soils, the space isoline map and soil type distribution map of Guiyang City were overlaid by using the software ArcView 3.2. The different level concentrations of heavy metals were statistical analyzed according to the different soil types. Meanwhile, to facilitate analysis, , soil types within the study area were divided into three types by human impact: man-influenced soil type, soil type by human impact in general and soil type with less human impact.

4.2.4.1 Soils types with stronger human impact: Yellow earths, limestone soils, paddy soils

Yellow soil lands an area of 3.335×10^5ha for the area of Guiyang City it totally lands 41.5% land area of Guiyang City. Yellow was born in hot and humid environmental conditions, rock weathering and fast weathering by strong leaching, base captions and silicon ions have leached, clay and the formation of secondary minerals constantly, ferric oxide relative aggregation, in which iron oxide by strong hydration, the formation of high water content of goethite ($Fe_2O \cdot H_2O$), limonite ($Fe_2O_3 \cdot 2H_2O$), more water, iron oxide ($Fe_2O.3H_2O$), the yellow hue of the minerals is the main hue(Sun,2002). Lime area 2.021×10^5ha, accounting for land area of Guiyang City, 25.2%. Guiyang Karst landforms, carbonate rocks are widely distributed, accounting for 80.63% area of Guiyang City, the corrosion - erosion, the erosion severe cases, the carbonate rocks exposed, weathering and limestone soil. Corrosion of carbonate karst weathering process is the release of Ca leaching, the residual calcium carbonate and clay minerals into the soil formed by the lime soil, shallow soil, and more with rock debris, soil properties affected by litho logy great, are rock soil. Limestone soil, with the abundant calcium and substitution of base level is high, the leaching process, due to constantly add calcium carbonate to the soil base to be preserved, weathering alteration of other minerals are also weak, delayed Al-rich off the role of silicon occur so long in a juvenile state. Distribution of lime in the soil and the soil zone boundaries clearly.

Paddy area is 1.156×10^5ha, accounting for the total land area of Guiyang City, 14.40%. In dam and hills in the valley bottom of the groove, light and heat conditions are good, as the irrigation and drainage conditions are good, the piece of the paddy field in the long-term hydroponics are the formation of the hydromorphic paddy soils that was the advanced stage of development of the paddy soil. As the periodic irrigation and drying, the soil, reduction and oxidation in alternating, the soil of iron, manganese and substances to restore migration, oxidative deposition, mind patterns of soil to form a brown rust, rust, and prism-like structure, which is a typical paddy soil types, is the main farming soil in Guiyang. The development at the initial stage of paddy rice soil infiltration education, are located in higher ground water table is low, almost no groundwater impact position. In the artificial irrigation, the irrigation of the soil affected by seasonal, alternating reduction and oxidation process, iron, manganese and base material was transported, deposited, in the former home territory, based on the formation of more than 20cm percogenic layer, initially with the characteristics of paddy soil. At the same time as soil pollution is not serious.

It shows that the levels of heavy metals were distributed at different levels, but mostly the first level and second level from Table 13 to Table 15.These three categories of soil type are the major soil types in Guiyang. By densely constraints, these three main soil types are local residents using the soil type. Yellow soil that the layer of soil and humus are thicker and soil acidity as well is the top soil for building timber forest and tea orchard. And its natural environment in which conditions is good, so it was open to most of the yellow land. Lime soil with high organic matter content, neutral to slightly alkaline, but the soil is thin, easy to dry; paddy soil water and heat conditions are better in the land after long-term aging and the formation of farming. In the long-term cultivation in the maturation process, human activities on soil heavy metal content of more. Therefore, the heavy metals content in different levels will have the distribution, while not a serious problem due to soil contamination, so in the first level and second level of the majority.

	Cd	Cr	Cu	Hg	Ni	Pb	Zn
1th level	73.2	20.2	74.2	96.7	37.2	63.7	68.4
2th level	21.1	61.1	18.9	1.5	45.4	29.0	23.4
3th level	3.7	16.5	4.1	0.5	12.3	4.8	4.6
4th level	1.4	1.9	1.8	0.3	4.0	1.7	1.7
5th level	0.4	0.3	0.5	0.2	0.8	0.5	1.9
6th level	0.2	0.0	0.5	0.8	0.3	0.3	0

Table 13. The percentage (%) of different concentration levels of heavy metals in Yellow earths of Guiyang.

	Cd	Cr	Cu	Hg	Ni	Pb	Zn
1th level	61.2	30.4	70.3	97	38.4	69.2	73.8
2th level	31.9	51.1	20.8	2.0	37.7	26.6	20.4
3th level	4.8	13.4	5.8	0.4	13.3	2.9	5.1
4th level	1.1	4.6	2.7	0.3	6.9	1	0.5
5th level	0.5	0.4	0.4	0.2	3.3	0.1	0.2
6th level	0.5	0.1	0	0.1	0.4	0.2	0

Table 14. The percentage (%) of different concentration levels of heavy metals in Limestone soils of Guiyang.

	Cd	Cr	Cu	Hg	Ni	Pb	Zn
1th level	69.6	22.6	74.2	96.8	40	68.3	72.6
2th level	21.7	58.2	17.4	2.3	39.9	28.1	18.1
3th level	6.3	16.7	4.9	0.4	13.7	2.6	6.5
4th level	1.6	2	1.5	0.4	4.7	0.8	1.5
5th level	0.3	0.5	1.5	0.1	1	0.1	1.3
6th level	0.5	0	0.5	0	0.7	0.1	0

Table 15. The percentage (%) of different concentration levels of heavy metals in Paddy soils of Guiyang.

4.2.4.2 Soils types with common human impact: Skeleton soils, purplish soils, litho soils, yellow-brown earths

Thick bone was 1.015×10^5 ha of 12.64% total land area in Guiyang city. Thick bone parent rocks were the weathering slope and residual consisting of shale and sand-shale. The soil body of soil type was instability, developed badly, thin soil and serious soil erosion. Phosphorus, potassium content is low. Purple soil area 104 ha, Guiyang 1.20 land area of 1.9%. Purple soil are mainly Jurassic purple red sandstone and mudstone tertiary surface soil after a. Purple rock type soft and crunchy, physical weathering speed, soil erosion is fast. And constantly weathering has added to make purple soil in the early stage of long-term. Purple clay mineral grains by weathering, silicon, carbonate etc iron compounds to form complex was stable in surface film, delaying the chemical weathering, keep the minerals of iron ore, the properties of soil siderite, thus presents. The rock soil due to constantly weathering of supplement, natural fertility soil of natural vegetation, also grew thick.

It can be seen in the heavy metals element content level mainly focus on four levels from Table 16 to Table 19. These four types of soil or natural conditions, using value is not high, Either the area is small, but the natural conditions, so the natural vegetation, soil for use, so not easily by the four types of human influence, but also affect the three types of great influence. Therefore, the four kinds of soil heavy metal content mainly concentrated in the top four.

	Cd	Cr	Cu	Hg	Ni	Pb	Zn
1th level	71.2	28.3	67.0	96	44.5	68	72.2
2th level	24.6	43.8	21.2	3.7	34.3	31.8	21.2
3th level	3.8	25.3	6.0	0.1	12.3	0.2	6.6
4th level	0.4	2.6	3.9	0.1	8.1	0	0
5th level	0	0	1.7	0	0.8	0	0
6th level	0	0	0.2	0.1	0	0	0

Table 16. The percentage (%) of different concentration levels of heavy metals in Skeletol soils.

	Cd	Cr	Cu	Hg	Ni	Pb	Zn
1th level	68.3	21.1	87.5	91.4	41.8	78.1	68.6
2th level	22.3	67.1	10.1	6.6	42.2	15.8	26.2
3th level	7.5	9.9	1.1	0.7	8.2	4.7	4.8
4th level	1.9	1.9	1.3	0.7	2.5	1.4	0.4
5th level	0	0	0	0.6	4.3	0	0
6th level	0	0	0	0	1.0	0	0

Table 17. The percentage (%) of different concentration levels of heavy metals in Purplish soils.

	Cd	Cr	Cu	Hg	Ni	Pb	Zn
1th level	51.3	37.7	51.4	98.9	21.4	78.5	66.0
2th level	39.3	38.0	33.1	1.1	42.7	18.5	30.6
3th level	7.8	11.4	11.1	0	16.9	1.4	3.2
4th level	1.5	7.0	4.4	0	6.2	1.6	0.2
5th level	0.1	5.6	0	0	5.1	0	0
6th level	0	0.3	0	0	7.7	0	0

Table 18. The percentage (%) of different concentration levels of heavy metals in Litho soils.

	Cd	Cr	Cu	Hg	Ni	Pb	Zn
1th level	52.3	28.5	90.0	79.3	45.2	41.7	78.6
2th level	42.8	71.1	10.0	13.8	41.5	21.0	16.7
3th level	4.9	0.4	0	3.1	7.3	25.1	0.7
4th level	0	0	0	2.2	2.7	10.3	1.4
5th level	0	0	0	1.3	3.3	0.4	2.6
6th level	0	0	0	0.3	0	1.5	0

Table 19. The percentage (%) of different concentration levels of heavy metals in Yellow-brown earths.

4.2.4.3 Soils types with less human impact: Mountain meadow soils and bog soils

In Guiyang City, The area of the Bog soils is 902ha, accounting for 0.11%of the total. On the surface of the local ground of the plateau, peal coal was formed by the accumulation of wet plants in the ancient swamp. Crustal movement in the later, ground up-list, swamp marsh broke off marsh gradually, black peat accumulated, the lower layers was white washing, which is the current peat.

Mountain meadow soil 379ha, accounting only 0.05% of land area of Guiyang. Mountain meadow soil is within the forest line, the gentle mountain top hi wet meadow and meadow shrub coppice Semis formed a class of soil. Such thin layers of soil, and generally contain gravel, grass surface layer with.

As shown in Table 20 to Table 21,the content of heavy metals in these two soil types are mainly concentrated in a Single interval, this phenomenon may be related to less human impact (such as fertilizer), heavy metals content in these soil types are mainly concentrated in the range corresponding their baseline values.

	Cd	Cr	Cu	Hg	Ni	Pb	Zn
1th level	0	16.8	100	0	100	0	100
2th level	100	83.2	0	100	0	0	0
3th level	0	0	0	0	0	17.4	0
4th level	0	0	0	0	0	82.6	0
5th level	0	0	0	0	0	0	0
6th level	0	0	0	0	0	0	0

Table 20. The percentage (%) of different concentration levels of heavy metals in Bog soils.

	Cd	Cr	Cu	Hg	Ni	Pb	Zn
1th level	100	65.8	98.2	100	0	0	100
2th level	0	34.2	1.8	0	100	100	0
3th level	0	0	0	0	0	0	0
4th level	0	0	0	0	0	0	0
5th level	0	0	0	0	0	0	0
6th level	0	0	0	0	0	0	0

Table 21. The percentage (%) of different concentration levels of heavy metals in Meadow solonchaks.

5. Conclusions

1. The mean value, standard deviation and maximum of As separately are 18.09 mg/kg, 11.57 mg/kg and 79.30 mg/kg in the surficial soil in Guiyang, of Cd separately are 0.302mg/kg, 0.363mg/kg, 2.620mg/kg and 95.7 percent smaller than 1.000mg/kg., of Cu separately are 43.1mg/kg, 30.3mg/kg, 213.0mg/kg and 94.6 percent smaller than 100.0mg/kg, of Pb are 43.2mg/kg, 31.3mg/kg, 318.9mg/kg and 95.5 per cent smaller than 100.0mg/kg, of Hg are 0.222mg/kg, 0.531mg/kg, 7.030mg/kg and 98.2 smaller

than 1.000mg/kg, of Ni are 38.3mg/kg, 14.9mg/kg, 102.5mg/kg and 95.8 percent smaller than 70.0mg/kg, of Zn are 84.7mg/kg, 49.8mg/kg, 385.0mg/kg and 94.3 percent smaller than 150.0mg/kg.

2. Comprehensively considering the results of calculation using the two kinds of methods, we respectively take 9.04mg/kg, 0.068 mg/kg, 18.8 mg/kg, 46.3 mg/kg, 14.8 mg/kg, 0.045 mg/kg, 44.0 mg/kg, 17.0 mg/kg as the baseline values of As, Cd, Cu, Zn, Pb, Hg, Cr and Ni in topsoil of Guiyang.

3. By the results of Geo-accumulation analysis of heavy metals in surficial soil, we get the surficial soil in 41 per cent of the Guiyang did not suffer the arsenic contaminative, In 40 per cent did not suffer the cadmium contaminative, In 46 per cent did not suffer the chromium contaminative, In 38 per cent did not suffer the copper contaminative, In 18 per cent did not suffer the lead contaminative, In 12 per cent did not suffer the mercury contaminative, In 19.2 per cent did not suffer the nickel contaminative.

4. By the results of contamination degree analysis, the maximum of arsenic contaminative degree of surficial soil in Guiyang is 4.27. Of cadmium are 12.1. Of chromium are 2.01. Of copper are 5.09. Of lead are 8.11. Of mercury are 45.87. Of nickel is 1.56. Of zinc are 2.85.

5. The soil types in this area were divided into three types of soil by the human impact degree. The three soil types of yellow soil, limestone soil and paddy soil that were the main soil types in Guiyang city were greatly influenced by human. The four soil types of skeleton soil, purple soil, stone soil and yellow brown soil that were not easily used were certainly influenced by human, the concentration of heavy metals in boggy soil and mountain meadow soil were concentrated on an interval, and tow types of soil (boggy soil and mountain meadow soil) were less influenced by human.

6. References

[1] Bauer I. and Bor J.1993.Vertikale Bilanzierung von Schwermetallen in Boden Kennzeichning der Empfinndlichkeit der boden gegenuber Schwermetallen unter Berucksichtigung von lithogenem Grundgehalt, pedogener An — und Abreicherung some antheopogener Zusatzbelastung, Teil 2. Texte56, Umweltbundesam, Berlin.

[2] Bauer I. and Bor J.1995.Lithogene, geonene und anthropogene Schwermetallgehalte von Lobboden an den Beispielen von Cu, Zn, Ni, Pb, Hg und Cd[J]. Mainzer Geowiss Mit. 24, 47-70

[3] Bauer I., Spernger M. and Bor J.1992.Die Berechnung Lithogener und geonerer Schwermetallgehalte von Lobboden am Beispielen von Cu, Zn und Pb[J]. Mainzer Geowiss Mitt. 21, 47-70

[4] Chen Cuihua, Ni Shijun, He Binbin, et al.,2008. Spatial-temporal variation of heavy metals contamination in sediments of the dexing mine, Jiangxi province[J]. Acta Geoscientica Sinica,29(5):639-646

[5] Chen Hangxin, Shen Xiachu, Yan Guangsheng. 1998. Research of experimental unit about international geochemical mapping. Edited by Wang Yanjun[M]. In The Geochemical paper of the 30th international geological meeting. Geological Press. Beijing, China.57-75

[6] Darnley A G · 1997. A global geochemical reference network: the foundation for geochemical baselines. J Geochemistry Exploration, 60(1):1-5

[7] Forstner U, Muller G. 1981.Concentrations of heavy metals and polycyclic aromatic hycarbons in river sediments: geochemical background, man's influence and environmental impact[J]. Geojournal, 5: 417~432.

[8] Forstner U, Ahlf W, Calmano W, et al. 1990.Sediment criteria development-contributions from environmental geochemistry to water quality management[A]. In: Heling D, Rothe P, Forstner U, et al. Sediments and environmental geochemistry: selected aspects and case histories[C]. Berlin Heidelberg:Springer-Verlag, 311~338.

[9] Hu Mianhao.2010. Application of index of geo-accumulation in evaluation of heavy metals pollution in Municipal sludge from Nanchang[J]. Guangdong Weiliang Yuansu Kexue,17(3):

[10] National Environmental Protect Bureau of China, National Technology Supervise Bureau of China.1995. GB15618-1995 Environmental Quality Standard for Soil [S].Beijing: Environmental Sciences Press of China(in Chinese)

[11] National Environmental Protect Bureau of China and National Technology Supervise Bureau of China.1998.GB/T17134-1997. Environmental Quality Determination of total arsenic-the silver diethyl dithiocarbamate photometric method[S]. Beijing: Environmental Sciences Press of China(in Chinese)

[12] National Environmental Protect Bureau of China and National Technology Supervise Bureau of China.1998.GB/T17136-1997. Environmental Quality Determination of total mercury-cold atomic absorption spectrophotometry method[S]. Beijing: Environmental Sciences Press of China(in Chinese)

[13] National Environmental Protect Bureau of China, National Technology Supervise Bureau of China. 1998. GB17137-1997.Environmental Quality Determination of total chromium-Flame atomic absorption spectrophotometry [S]. Beijing: Environmental Sciences Press of China(in Chinese)

[14] National Environmental Protect Bureau of China, National Technology Supervise Bureau of China. 1998. GB17138-1997.Environmental Quality Determination of Cu and Zn-Flame atomic absorption spectrophotometry [S]. Beijing: Environmental Sciences Press of China(in Chinese)

[15] National Environmental Protect Bureau of China, National Technology Supervise Bureau of China. 1998. GB17139-1997.Environmental Quality Determination of Ni-Flame atomic absorption spectrophotometry [S]. Beijing: Environmental Sciences Press of China(in Chinese)

[16] National Environmental Protect Bureau of China and National Technology Supervise Bureau of China.1998.GB/T17141-1997.Soil Quality-Determination of Lead and Cadmium: Graphite Furnace Atomic Absorption Spectrophotometry[S]. Beijing: Environmental Sciences Press of China(in Chinese)

[17] National Environmental Protect Bureau of China, 2004. HJ/T 166-2004. The technical specification for soil environmental monitoring[S].Beijing Environmental Sciences Press of China(in Chinese)

[18] Lepeltier C. 1969. A simplified treatment of geochemical data by graphical represatation[J]. Enviromental Geology, 64, 538-550

[19] Liao Zi Ji.1992. Environmental chemistry and biological effects of trace elements[M]. BeiJing. China Environmental Science Press, 178-253

[20] Li Qiang, Mo Dalun.1997. The damage and research progress of As contamination in the soil environment [J]. Tropical and Subtropical Soil Science, 6(4): 291 - 295

[21] Matschullat J., Ottenstein R. and Reimann C. (2000) Geochemical background-can we calculate it[J] Environmental Geology, 39, 990-1000

[22] Muller G. 1969, Index of geoaccumulation in sediments of the Rhine River[J]. Geojournal, 2(3): 108~118
[23] Salminen R, Tarvainen T, 1997. The problem of defining geochemical baseline. A case study of selected elements and geological materials in Finland. J. Geochemical Exploration, 60(1): 91-98
[24] Salminen R, Gregorauskiene V. 2000. Consideration regarding the definition of a geochemical baseline of elements in the surfical materials in areas differing in basic geology. Applied Geochemistry, 15: 647-653
[25] Sun Chengxing.2002.Red weathering material sources and rare earth elements chemical research in Karst area of Guizhou[D]
[26] Teng Yanguo, Ni Shijun, Tuo Xianguo et al., 2002.Geochemical baseline and trace metal pollution of soil in Panzhihua mining area[J]. Chinese Journal of Geochemistry. 21, 274-281
[27] Teng Yanguo, Tuo Xianguo, Ni Shijun et al., 2003.Environment geochemical of heavy metal contaminants in soil and stream sediment in Panzhihua mining and smelting area, southwestern China[J]. Chinese Journal of Geochemistry, 22, 254-262
[28] Xu Hong-ning, Xu Jia-lin.1996. The cause and the distribution of As abnormal area in China [J]. Soil, 2:80-84
[29] Yi Xiu, Gu Xiaojing, Hou Yanqing, et al., 2010. Assessment on soil heavy metals pollution by Geo-accumulation index in Jinghuiqu irrigation district of Shaanxi province[J]. Journal of Earth Sciences and Environment,32(3):288-291

Permissions

The contributors of this book come from diverse backgrounds, making this book a truly international effort. This book will bring forth new frontiers with its revolutionizing research information and detailed analysis of the nascent developments around the world.

We would like to thank Dr. Dionisios Panagiotaras, for lending his expertise to make the book truly unique. He has played a crucial role in the development of this book. Without his invaluable contribution this book wouldn't have been possible. He has made vital efforts to compile up to date information on the varied aspects of this subject to make this book a valuable addition to the collection of many professionals and students.

This book was conceptualized with the vision of imparting up-to-date information and advanced data in this field. To ensure the same, a matchless editorial board was set up. Every individual on the board went through rigorous rounds of assessment to prove their worth. After which they invested a large part of their time researching and compiling the most relevant data for our readers. Conferences and sessions were held from time to time between the editorial board and the contributing authors to present the data in the most comprehensible form. The editorial team has worked tirelessly to provide valuable and valid information to help people across the globe.

Every chapter published in this book has been scrutinized by our experts. Their significance has been extensively debated. The topics covered herein carry significant findings which will fuel the growth of the discipline. They may even be implemented as practical applications or may be referred to as a beginning point for another development. Chapters in this book were first published by InTech; hereby published with permission under the Creative Commons Attribution License or equivalent.

The editorial board has been involved in producing this book since its inception. They have spent rigorous hours researching and exploring the diverse topics which have resulted in the successful publishing of this book. They have passed on their knowledge of decades through this book. To expedite this challenging task, the publisher supported the team at every step. A small team of assistant editors was also appointed to further simplify the editing procedure and attain best results for the readers.

Our editorial team has been hand-picked from every corner of the world. Their multi-ethnicity adds dynamic inputs to the discussions which result in innovative outcomes. These outcomes are then further discussed with the researchers and contributors who give their valuable feedback and opinion regarding the same. The feedback is then collaborated with the researches and they are edited in a comprehensive manner to aid the understanding of the subject.

Apart from the editorial board, the designing team has also invested a significant amount of their time in understanding the subject and creating the most relevant covers. They scrutinized every image to scout for the most suitable representation of the subject and create an appropriate cover for the book.

The publishing team has been involved in this book since its early stages. They were actively engaged in every process, be it collecting the data, connecting with the contributors or procuring relevant information. The team has been an ardent support to the editorial, designing and production team. Their endless efforts to recruit the best for this project, has resulted in the accomplishment of this book. They are a veteran in the field of academics and their pool of knowledge is as vast as their experience in printing. Their expertise and guidance has proved useful at every step. Their uncompromising quality standards have made this book an exceptional effort. Their encouragement from time to time has been an inspiration for everyone.

The publisher and the editorial board hope that this book will prove to be a valuable piece of knowledge for researchers, students, practitioners and scholars across the globe.

List of Contributors

Paula Álvarez-Iglesias and Belén Rubio
Universidad de Vigo, Vigo (Pontevedra), Spain

Nikos Nanos
School of Forest Engineering - Madrid Technical University, Ciudad Universitaria s/n, Madrid, Spain

José Antonio Rodríguez Martín
I.N.I.A. Department of the Environment, Madrid, Spain

Orce Spasovski
University "Goce Delcev" Stip, Faculty of Natural and Technical Sciences, Macedonia

Leyla Kalender
Firat University, Department of Geologycal Engineering, Elazig, Turkey

Yu N. Vodyanitskii and A. T. Savichev
Department of Soil Science, Moscow State University; Geological Institute, RAS, Russia

Tatyana Gunicheva
A. P. Vinogradov Institute of Geochemistry, Russia

B. Anjan Kumar Prusty, Rachna Chandra and P. A. Azeez
Environmental Impact Assessment Division, Sálim Ali Centre for Ornithology and Natural History (SACON), Anaikatty (PO), Coimbatore, India

Arthur James Swart
Vaal University of Technology, Republic of South Africa

S. G. Viladkar
Carbonatite Research Centre, Amba Dongar, Kadipani, District Vadodara, India

Ji Wang and Yixiu Zhang
School of Geographical and Environment Sciences, Guizhou Normal University, Guiyang, Guizhou, China
Key Laboratory of Remote Sensing Applications in Resources and Environments, Guizhou, China

Printed in the USA
CPSIA information can be obtained
at www.ICGtesting.com
JSHW011428221024
72173JS00004B/721